Hydrogen, Presented By Hildy, From The Magical Elements of the Periodic Table Book Series

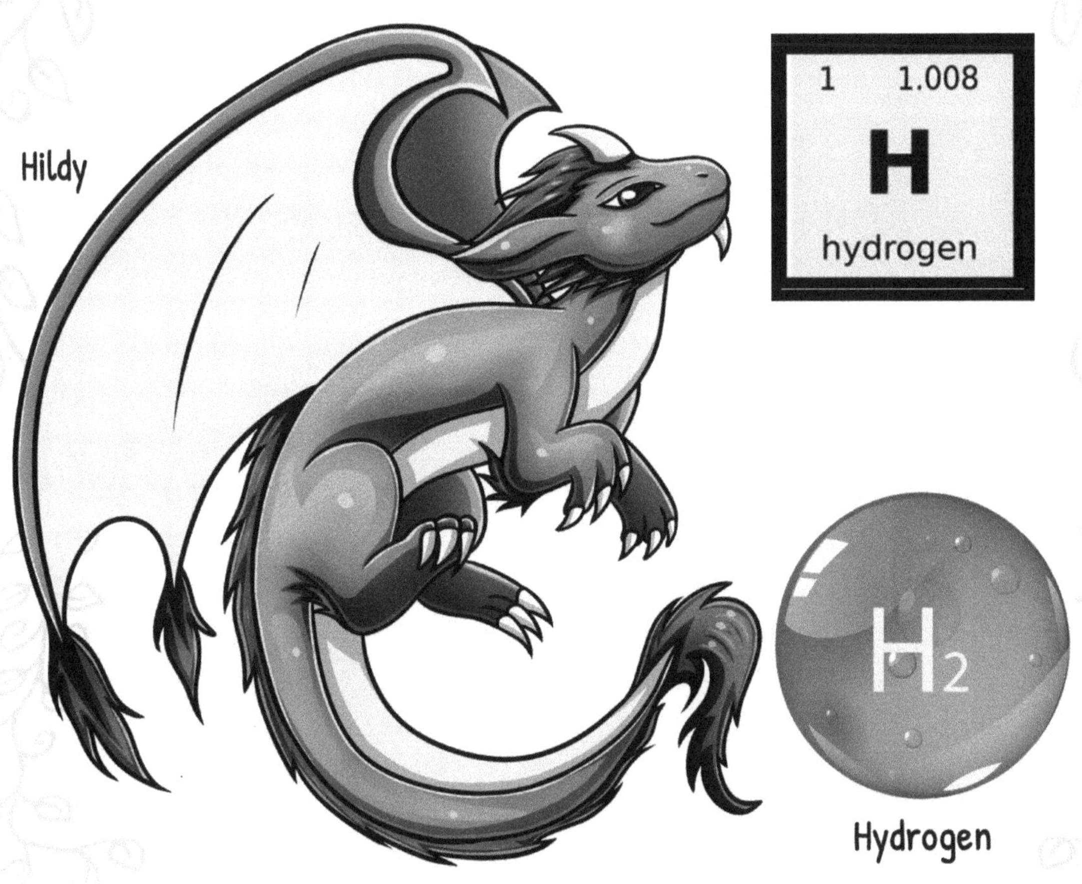

By Sybrina Durant with Illustrations by Pranavva et al.

Hydrogen, Presented By Hildy, From The Magical Elements of the Periodic Table Book Series

Story copyright 2026

KDP Soft Cover ISBN: 9798195289782

BISAC Codes:

JNF051070 JUVENILE NONFICTION / Science & Nature / Chemistry

JNF016000 JUVENILE NONFICTION / Curiosities & Wonders

JNF051080 JUVENILE NONFICTION / Science & Nature / Earth Sciences / General

Soft Cover Print ISBN 13 — 978-1-942740-66-7

Hildy Presents Hydrogen

This Element 1 book features the periodic table element, Hydrogen. It is presented by Hildy, a member of the Elemental Dragon Clan. Each dragon has a magical tail tipped with an element that gives them unique powers. Their powers are based on the properties of their element.

Hildy is just one of the 118 elementals who will present all of the Magical Elements of the Periodic Table to readers who are curious about the wonders of the world.

Hildy introduces the very magical element, Hydrogen, in her book.

The Elemental Dragon Clan and their other techno-magical friends are the perfect group to introduce you to metals and other elements in the Periodic Table. Hopefully, this Magical Element of the periodic table book will spark an interest in the magical and real world properties of all the metals and other elements known today. You may be surprised at how prominently they feature in our every day lives.

Each page in this book contains terms that might not be completely familiar to the reader. Refer to the definitions in the back of the book to get a clear understanding of each meaning.

There is also a fun elemental themed Periodic Table at the back of the book. It features 118 elements presented by fanciful characters like unicorns, dragons, wizards, knights and goblins.. They want you to remember that if there's no metal...there's no magic or technology.

Remember, "No metal – No Magic. . .and No Technology".

It's Techo-Magical.

Note: Sybrina Publishing websites are Sybrina.com and MagicalPTElements.com. Follow sybrinapublishing on Instagram, Magical Elements of the Periodic Table on Facebook, @sybrinad on Pinterest, Sybrina_SPT on Twitter; and Sybrina Durant on LinkedIn.

Hydrogen is a Non-Metal

- In the 1500s, Paracelsus observed bubbles from iron and sulfuric acid mixtures in the mining districts of Austria but didn't identify it as a new substance. In 1671, Robert Boyle produced hydrogen by mixing iron filings with dilute acids. Later, in 1766, Henry Cavendish recognized it as a distinct element, calling it "inflammable air."

- Hydrogen is the lightest gas and element. It is about 14 times lighter than air, so it rises easily. It is a diatomic gas with no color or smell, and it can catch fire easily.

- Hydrogen gas does not carry heat or electricity well because it is a non-metal and has no free charged particles. In this form, it acts like an insulator.

- Hydrogen is very weakly diamagnetic, meaning it is weakly repelled by magnetic fields.

- Hydrogen is classified as a Non-Metal, not a noble gas, because it has only 1 electron in its outer shell and needs 1 more to fill it. Noble gases already have full outer shells. Hydrogen is a reactive gas that exists as two atoms joined together. It makes covalent bonds, does not conduct electricity well, and has a high ionization energy. These are all traits shared by non-metals.

LEGEND

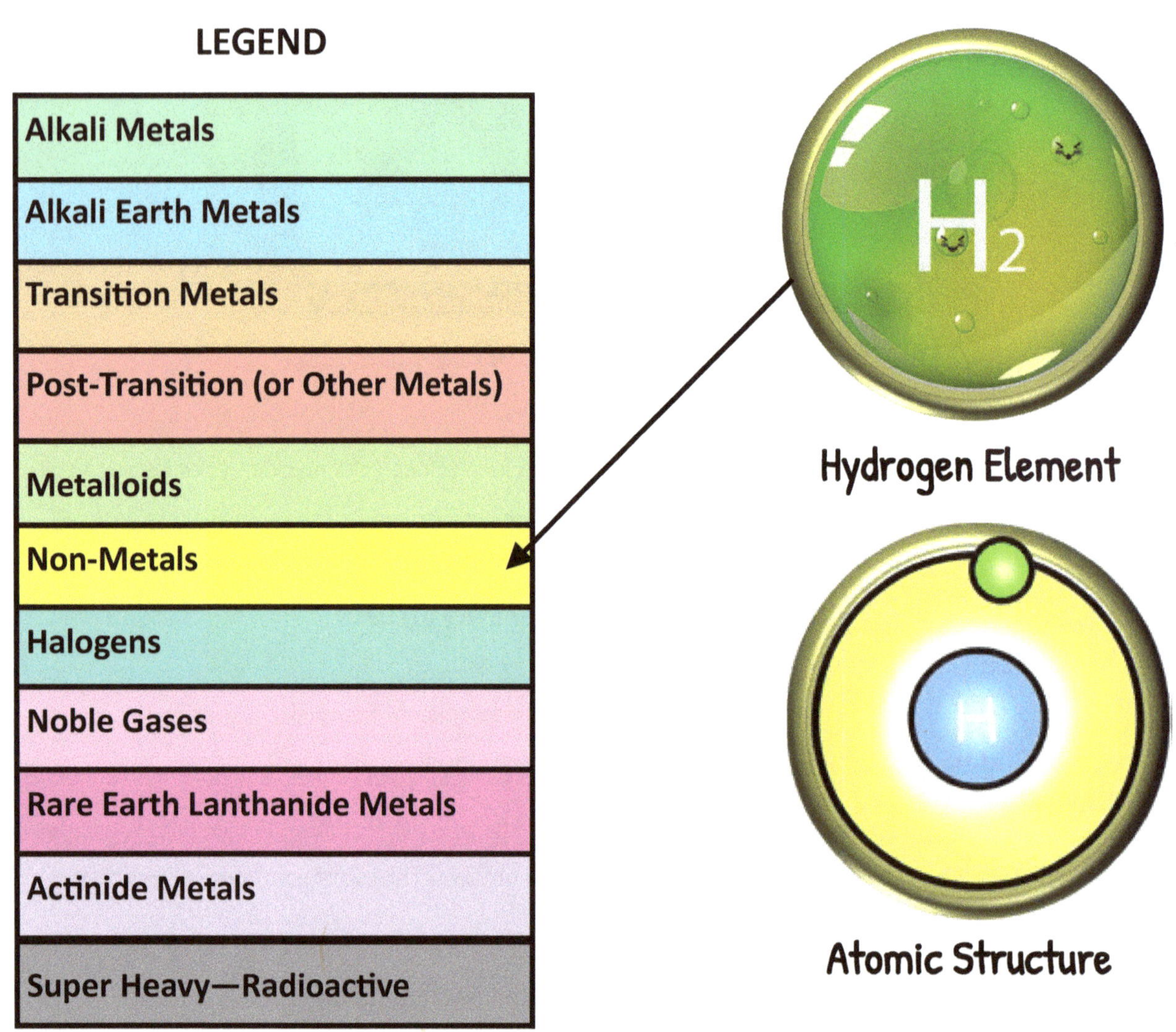

Hydrogen Element

Atomic Structure

Non-Metals—These elements reside in columns 15-17, and can be gases, liquids, or solids. They don't conduct heat or electricity. The solids are brittle, and they have no metallic luster. They readily accept electrons from metals to form salts. These include nitrogen, oxygen, fluorine, chlorine, bromine, and iodine.

Hydrogen is the lightest element and one of the most common elements in the universe. People have been interested in it for a long time because it can burn easily and it is very light. Since it was first discovered in the 1500s, hydrogen has been used in several different ways. Some of these early uses were important at the time, but over the years hydrogen was mostly replaced by safer, cheaper, or easier options.

One of the first names for hydrogen was "inflammable air." In 1766, the British scientist Henry Cavendish separated it and found that it burned very easily. This made people curious about it, and they began testing how it could be used. One early use was for balloon flight. Because hydrogen is so light, it could help balloons rise into the air. In 1783, the Montgolfier brothers used hydrogen in the first successful manned balloon flight, and this opened the door to early air travel experiments.

Hydrogen was also used for lighting in the 1800s. It was put into gas lamps because it gave off a bright flame. At the time, this was useful because it was cleaner than candles or oil lamps. Later, hydrogen was used in welding and cutting metal. Oxy-hydrogen torches could create very high heat, which made them useful in some factory and metalworking jobs. Hydrogen was also used in airships during the early 1900s because it could lift them into the sky. But this use became far less common after the Hindenburg disaster in 1937, when a hydrogen-filled airship caught fire and crashed. After that, many people became afraid of hydrogen, and safer gases like helium were used instead.

Hydrogen is not widely used for these old purposes today for several reasons. One big reason is safety. Hydrogen can catch fire very easily, and the Hindenburg crash showed how dangerous it can be if it is not handled carefully. Another reason is that other choices have become better. Electric lights replaced gas lighting because they were safer, brighter, and easier to use. Other fuels like acetylene and propane replaced hydrogen in welding because they were more practical in many work settings. Hydrogen is also hard to store and transport because it is a very light gas and takes up a lot of space unless it is compressed or made very cold, which adds cost and complexity.

Even though hydrogen is no longer used much for those older jobs, it is being looked at again as a clean energy source for the future. This is because hydrogen can be made in ways that do not produce as much pollution, especially when it is made using renewable electricity. When hydrogen is used, it can produce energy without releasing carbon dioxide at the point of use, which makes it very interesting for fighting climate change. Many people now see it as useful for transport, energy storage, and factory work.

In transport, hydrogen may help power cars, buses, trucks, trains, ships, and even some aircraft in the future. Hydrogen fuel cells can turn hydrogen into electricity, which can run electric motors. This can give vehicles long driving ranges and fast refueling times, which is helpful for large vehicles and long trips. In energy storage, hydrogen can store extra power from wind and solar energy. When there is too much electricity being made, it can be used to produce hydrogen. Later, that hydrogen can be turned back into energy when needed, helping balance the power grid. In factories, hydrogen may help replace coal, oil, and natural gas in some high-heat jobs. This is especially important in industries like steel, cement, and chemicals, where lowering pollution is very hard but very important.

So, while hydrogen was first used for things like balloons, lighting, welding, and airships, it is now being seen in a new way. Instead of being mainly a gas for old technologies, it may become an important part of cleaner energy systems in the future.

Uses For Hydrogen

Hydrogen is used to power vehicles in some countries. It can be turned into energy that helps cars, buses, and trucks move. Some people like hydrogen because it makes less pollution than gas or diesel. It can also be filled into vehicles fairly quickly. However, not many places have enough stations for it yet. In some countries, hydrogen is seen as a clean fuel for the future and a possible way to reduce harm to the environment.

Hydrogen is used in hydrocracking, a process that helps turn heavy refinery products, like crude oil, into more useful fuels and chemicals. In this process, the large molecules are broken, or "cracked," into smaller ones. This makes products such as gasoline and diesel. It also helps create many other important chemicals used in everyday life. Hydrocracking is a useful way to get more value from heavy oil materials.

Uses For Hydrogen

(Continued)

Hydrogen is used in a process called hydrogenation. This process changes unsaturated fats into more saturated fats and oils. Food makers use it to make hydrogenated vegetable oils. These oils are found in many foods, such as margarine and peanut butter. Hydrogenation helps change the texture and shelf life of these products. Hydrogen is also used in some skin care products. It helps make certain creams and lotions smoother and more stable. In this way, hydrogen has many uses in both food and personal care products.

Next to oil refineries, ammonia is now the biggest use of hydrogen. Ammonia, or NH3, is used to make ammonium nitrate, which is a common fertilizer that helps crops grow. It is also found in many cleaning products used around the home. Because ammonia can be strong and can irritate skin, it is important to handle it carefully. Always wear gloves when using products that contain ammonia. This helps protect your hands and keeps you safe while cleaning or working with chemicals. Ammonia plays a useful role in farming and cleaning, but it should always be used with care.

The Source of Hydrogen

Green Hydrogen is produced by using electricity to split water molecules (H_2O) into hydrogen (H_2) and oxygen (O_2). Blue hydrogen is produced from hydrocarbon sources, such as natural gas, and combined with carbon capture technology.

Hydrogen is the lightest and most common element in the universe. It is used in energy, transport, factories, and science. More people are talking about hydrogen now because many countries want cleaner energy. But one question is very important: where does hydrogen come from?

The short answer is that hydrogen is not usually found alone in nature. Most of the time, it is joined with other elements. For example, it is found in water, which is made of hydrogen and oxygen. It is also found in fuels like natural gas, oil, and coal. So, before people can use hydrogen, they must separate it from these other materials.

One of the main ways to make hydrogen is called steam methane reforming, or SMR. This is used in many countries. In this process, natural gas is mixed with very hot steam. The heat causes a chemical change that creates hydrogen and carbon dioxide. This method is popular because it can make a lot of hydrogen and it is cheaper than some other methods. The problem is that it also makes carbon dioxide, which is a gas that adds to climate change.

Natural gas comes from deep underground. It is a fossil fuel that formed millions of years ago from dead plants and animals buried under rock and soil. People get it by drilling into the ground. After it is brought up, it can be used for heating, cooking, making electricity, or making hydrogen. Because it is easy to use and move, it is one of the main sources for hydrogen production.

The Source of Hydrogen (Continued)

Another way to make hydrogen is electrolysis. In this process, electricity is used to split water into hydrogen and oxygen. Water is put into a machine called an electrolyzer. Then electricity passes through the water and breaks it apart. The hydrogen gas is collected and used.

Electrolysis is a very useful method because it can make hydrogen from water, which is common. It can also be much cleaner if the electricity comes from wind, solar, or water power. But if the electricity comes from coal or natural gas, then the process is not as clean.

Hydrogen can also come from biomass gasification. Biomass means natural material from plants and animals. This often includes crop waste, wood, grass, and other plant matter. Gasification means heating this material in a special way so it turns into gas. That gas can then be processed to make hydrogen.

This method is useful because it can use waste materials that might otherwise be thrown away. It can also reduce the need for fossil fuels. However, it must be managed carefully so it does not cause too much pollution or use land in harmful ways.

In some factories, hydrogen is made as a byproduct. This means it is not the main thing being made, but it is produced during another process. Instead of wasting it, companies can collect and use it. This helps reduce waste and makes better use of resources.

After hydrogen is made, it often needs to be cleaned. Raw hydrogen may have other gases mixed in, such as carbon dioxide, water vapor, sulfur, or leftover chemicals. These must be removed before the hydrogen can be used well. Some uses, like fuel cells and science work, need very pure hydrogen.

Two common cleaning methods are pressure-swing adsorption, or PSA, and membrane separation. These systems remove unwanted gases and leave cleaner hydrogen behind. This step is important because even small amounts of dirt can affect performance and safety.

China is one of the biggest hydrogen producers in the world. This is because it has a huge industrial economy and needs lots of energy. China makes hydrogen using coal gasification, steam methane reforming, and electrolysis. Japan and the United States also produce a lot of hydrogen for industry, research, and clean energy plans.

The effect of hydrogen on the environment depends on how it is made. Hydrogen made from fossil fuels creates a lot of carbon pollution. This is often called gray hydrogen. If the carbon is captured and stored, it may be called blue hydrogen. If hydrogen is made with renewable electricity, it is called green hydrogen. Green hydrogen is the cleanest type, but it is often more expensive.

The way fossil fuels are taken from the ground can also harm the environment. Drilling for natural gas can cause methane leaks. Methane is a very strong greenhouse gas. Transporting and processing fossil fuels can also cause pollution. So the full effect of hydrogen depends not just on the gas itself, but on how it was made.

Hydrogen is more than just a gas. It is part of a much bigger energy system. It can help power homes, factories, vehicles, and future clean energy projects. Governments, companies, and scientists are working on better ways to make hydrogen cleaner, cheaper, and safer. They are also studying how to store it, move it, and use it well.

In the future, hydrogen may become even more important. It could help cut carbon pollution in heavy transport, steel making, shipping, and energy storage. But for that to happen, we need better ways to produce it.

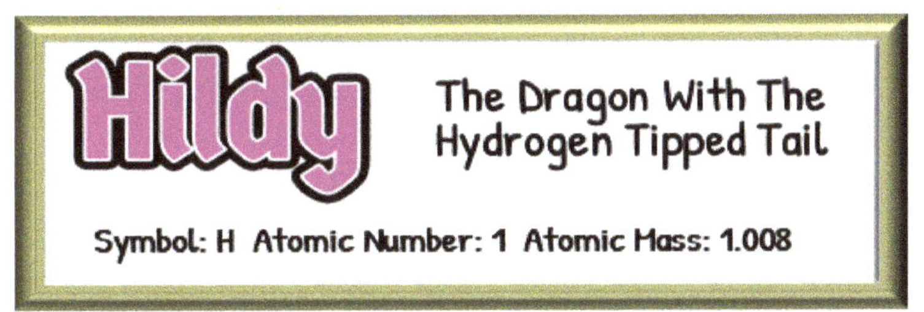

Hydrogen resides in Group 1 Period 1 on the Periodic Table.

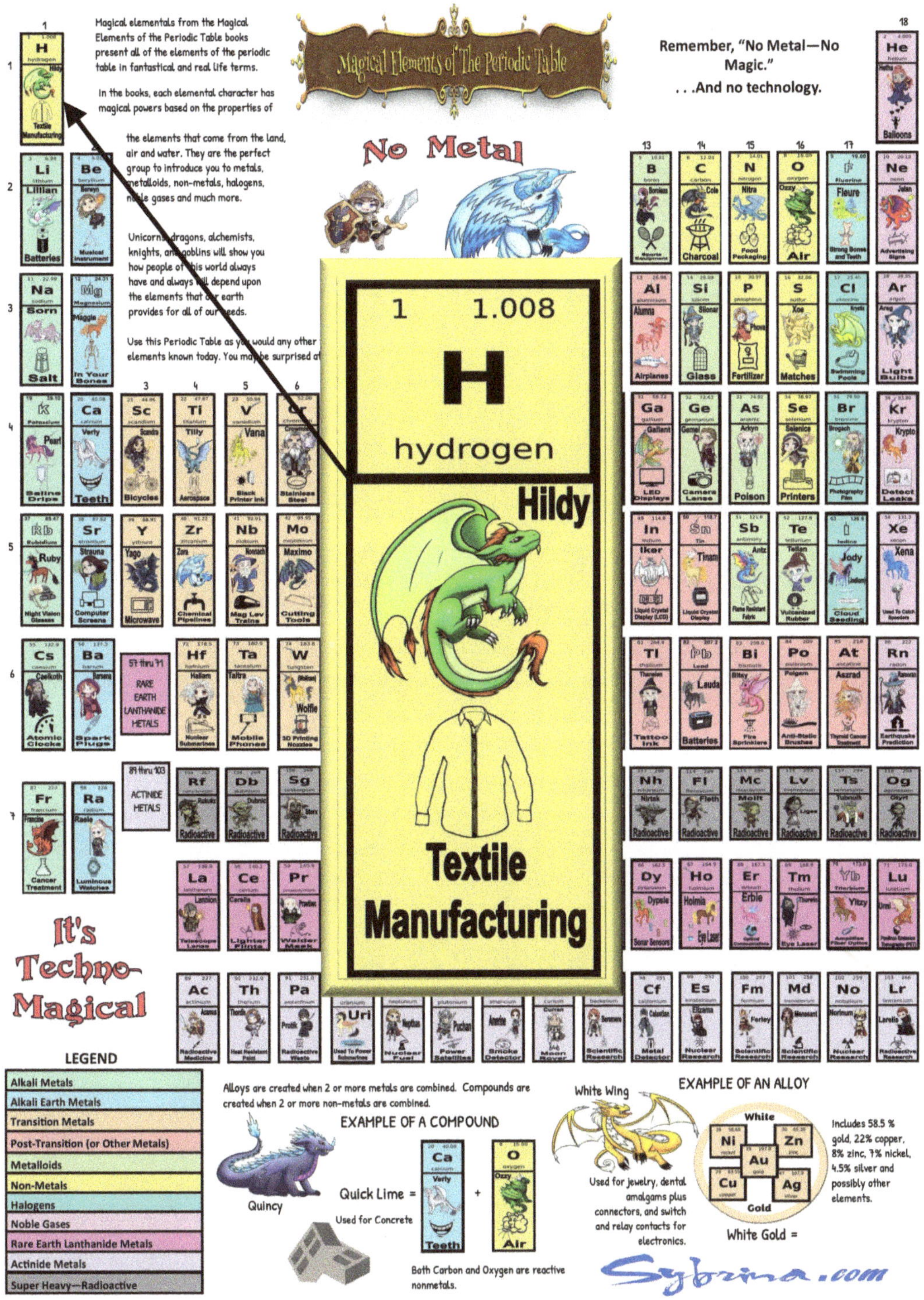

Magical elementals from the Magical Elements of the Periodic Table books present all of the elements of the periodic table in fantastical and real life terms.

In the books, each elemental character has magical powers based on the properties of the elements that come from the land, air and water. They are the perfect group to introduce you to metals, metalloids, non-metals, halogens, noble gases and much more.

Unicorns, dragons, alchemists, knights, and goblins will show you how people of this world always have and always will depend upon the elements that our earth provides for all of our needs.

Use this Periodic Table as you would any other elements known today. You may be surprised at

Remember, "No Metal—No Magic."
. . .And no technology.

Magical Elements of The Periodic Table

No Metal

It's Techno-Magical

LEGEND

| Alkali Metals |
| Alkali Earth Metals |
| Transition Metals |
| Post-Transition (or Other Metals) |
| Metalloids |
| Non-Metals |
| Halogens |
| Noble Gases |
| Rare Earth Lanthanide Metals |
| Actinide Metals |
| Super Heavy—Radioactive |

Alloys are created when 2 or more metals are combined. Compounds are created when 2 or more non-metals are combined.

EXAMPLE OF A COMPOUND

Quincy

Quick Lime = Ca + O

Used for Concrete

Both Carbon and Oxygen are reactive nonmetals.

White Wing

Used for jewelry, dental amalgams plus connectors, and switch and relay contacts for electronics.

EXAMPLE OF AN ALLOY

White Gold =

Includes 58.5 % gold, 22% copper, 8% zinc, 7% nickel, 4.5% silver and possibly other elements.

Sybrina.com

Magical Elements of The Periodic Table
Magical elementals from the Magical Elements of the Periodic Table books present all of the elements of the periodic table in fantastical and real life terms.
In the books, each elemental character has magical powers based on the properties of the elements that come from the land, air and water. They are the perfect group to introduce you to metals, metalloids, non-metals, halogens, noble gases and much more.
Unicorns, dragons, alchemists, knights, and goblins will show you how people of this world always have and always will depend upon the elements that our earth provides for all of our needs.
Use this Periodic Table as you would any other to spark an interest in the magical and real world properties of all the elements known today. You may be surprised at how prominently they feature in our every day lives.
No Metal
Actinium To Zirconium
No Magic
Remember, "No Metal— No Magic." . . .And no technology.

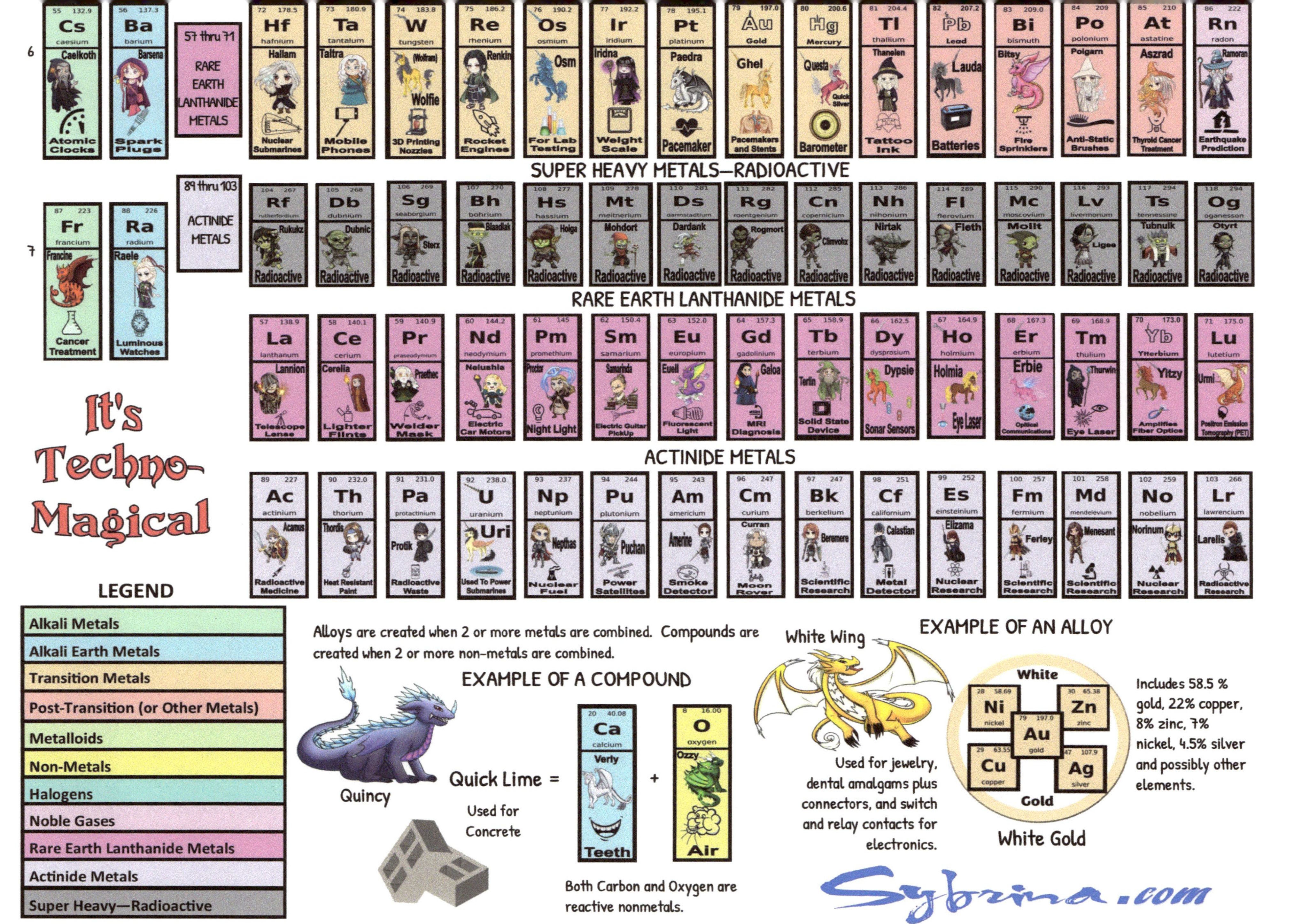

It's Techno-Magical

55 132.9 Cs caesium Caelkoth Atomic Clocks
56 137.3 Ba barium Barsena Spark Plugs
6
57 thru 71 RARE EARTH LANTHANIDE METALS
72 178.5 Hf hafnium Hallam Nuclear Submarines
73 180.9 Ta tantalum Taltra Mobile Phones
74 183.8 W tungsten (Wolfram) Wolfie 3D Printing Nozzles
75 186.2 Re rhenium Renkin Rocket Engines
76 190.2 Os osmium Osm For Lab Testing
77 192.2 Ir iridium Iridna Weight Scale
78 195.1 Pt platinum Paedra Pacemaker
79 197.0 Au gold Ghel Pacemakers and Stents
80 200.6 Hg mercury Questa Quick Silver Barometer
81 204.4 Tl thallium Thanelen Tattoo Ink
82 207.2 Pb lead Lauda Batteries
83 209.0 Bi bismuth Bitsy Fire Sprinklers
84 209 Po polonium Polgarn Anti-Static Brushes
85 210 At astatine Aszrad Thyroid Cancer Treatment
86 222 Rn radon Ramoran Earthquake Prediction

7
87 223 Fr francium Francine Cancer Treatment
88 226 Ra radium Raele Luminous Watches
89 thru 103 ACTINIDE METALS

SUPER HEAVY METALS—RADIOACTIVE
104 267 Rf rutherfordium Rukukz Radioactive
105 268 Db dubnium Dubnic Radioactive
106 269 Sg seaborgium Sterx Radioactive
107 270 Bh bohrium Blaadlak Radioactive
108 277 Hs hassium Hoiga Radioactive
109 278 Mt meitnerium Mohdort Radioactive
110 281 Ds darmstadtium Dardank Radioactive
111 282 Rg roentgenium Rogmort Radioactive
112 285 Cn copernicium Clinvolz Radioactive
113 286 Nh nihonium Nirtak Radioactive
114 289 Fl flerovium Fleth Radioactive
115 290 Mc moscovium Molit Radioactive
116 293 Lv livermorium Ligee Radioactive
117 294 Ts tennessine Tubnulk Radioactive
118 294 Og oganesson Otyrt Radioactive

RARE EARTH LANTHANIDE METALS
57 138.9 La lanthanum Lannion Telescope Lense
58 140.1 Ce cerium Cerelia Lighter Flints
59 140.9 Pr praseodymium Praethec Welder Mask
60 144.2 Nd neodymium Nelushla Electric Car Motors
61 145 Pm promethium Proctor Night Light
62 150.4 Sm samarium Samarlda Electric Guitar PickUp
63 152.0 Eu europium Euell Fluorescent Light
64 157.3 Gd gadolinium Galoa MRI Diagnosis
65 158.9 Tb terbium Terlin Solid State Device
66 162.5 Dy dysprosium Dypsie Sonar Sensors
67 164.9 Ho holmium Holmia Eye Laser
68 167.3 Er erbium Erbie Optical Communications
69 168.9 Tm thulium Thurwin Eye Laser
70 173.0 Yb Ytterbium Yitzy Amplifies Fiber Optics
71 175.0 Lu lutetium Urmi Positron Emission Tomography (PET)

ACTINIDE METALS
89 227 Ac actinium Acamus Radioactive Medicine
90 232.0 Th thorium Thordis Heat Resistant Paint
91 231.0 Pa protactinium Protik Radioactive Waste
92 238.0 U uranium Uri Used To Power Submarines
93 237 Np neptunium Nepthas Nuclear Fuel
94 244 Pu plutonium Puchan Power Satellites
95 243 Am americium Amerine Smoke Detector
96 247 Cm curium Curran Moon Rover
97 247 Bk berkelium Beremere Scientific Research
98 251 Cf californium Calastian Metal Detector
99 252 Es einsteinium Elizama Nuclear Research
100 257 Fm fermium Ferley Scientific Research
101 258 Md mendelevium Menesant Scientific Research
102 259 No nobelium Norinum Nuclear Research
103 266 Lr lawrencium Larelis Radioactive Research

LEGEND
Alkali Metals
Alkali Earth Metals
Transition Metals
Post-Transition (or Other Metals)
Metalloids
Non-Metals
Halogens
Noble Gases
Rare Earth Lanthanide Metals
Actinide Metals
Super Heavy—Radioactive

Alloys are created when 2 or more metals are combined. Compounds are created when 2 or more non-metals are combined.

EXAMPLE OF A COMPOUND
Quincy
Quick Lime =
Used for Concrete
20 40.08 Ca calcium Verly Teeth
+
8 16.00 O oxygen Ozzy Air
Both Carbon and Oxygen are reactive nonmetals.

White Wing
EXAMPLE OF AN ALLOY
Used for jewelry, dental amalgams plus connectors, and switch and relay contacts for electronics.
White
28 58.69 Ni nickel
30 65.38 Zn zinc
79 197.0 Au gold
29 63.55 Cu copper
47 107.9 Ag silver
Gold
White Gold
Includes 58.5 % gold, 22% copper, 8% zinc, 7% nickel, 4.5% silver and possibly other elements.

Sybrina.com

All Of The Periodic Table Elements Listed Alphabetically

Element Listed In Red Is Featured In This Book

ACTINIUM—*AC*—89

ALUMINUM—*AL*—13

AMERICIUM—*AM*—95

ANTIMONY—*SB*—51

ARGON—*AR*—18

ARSENIC—*AS*—33

ASTATINE—*AT*—85

BARIUM—*BA*—56

BERKELIUM—*BK*—97

BERYLLIUM—*BE*—4

BISMUTH—*BI*—83

BOHRIUM—*BH*—107

BORON—*B*—5

BROMINE—*BR*—35

CADMIUM—*CD*—48

CALCIUM (Vital)—*CA*—20

CALIFORNIUM—*CF*—98

CARBON—*C*—6

CERIUM—*CE*—58

CESIUM—*CS*—55

CHLORINE (Keen)—*CL*—17

CHROMIUM—*CR*—24

COBALT—*CO*—27

COPERNICIUM—*CN*—112

COPPER—*CU*—29

CURIUM—*CM*—96

DARMSTADTIUM—*DS*—110

DUBNIUM—*DB*—105

DYSPROSIUM—*DY*—66

ERBIUM—*ER*—68

EINSTEINIUM—*ES*—99

EUROPIUM—*EU*—63

FERMIUM—*FM*—100

FLEROVIUM—*FL*—114

FLUORINE—*F*—9

FRANCIUM—*FR*—87

GADOLINIUM—*GD*—64

GALLIUM—*GA*—31

GERMANIUM—*GE*—32

GOLD—*AU*—79

HAFNIUM—*HF*—72

HASSIUM—*HS*—108

HELIUM—*HE*—2

HOLMIUM—*HO*—67

HYDROGEN—*H*—1

INDIUM—*IN*—49

IODINE (JODIUM) —*I*—53

IRIDIUM—*IR*—77

IRON—*FE*—26

KRYPTON—*KR*—36

LANTHANUM—*LA*—57

LAWRENCIUM—*LR*—103

LEAD—*PB*—82

LITHIUM—*LI*—3

LIVERMORIUM—*LV*—116

LUTETIUM (Unique)—*LU*—71

MAGNESIUM—*MG*—12

MANGANESE—*MN*—25

MEITNERIUM—*MT*—109

MENDELEVIUM—*MD*—101

MERCURY (QUICK SILVER) —*HG*—80

MOLYBDENUM—*MO*—42

MOSCOVIUM—*MC*—115

NEODYMIUM—*ND*—60

NEON (Jazzy)—*NE*—10

NEPTUNIUM—*NP*—93

NICKEL—*NI*—28

NIHONIUM—*NH*—113

NIOBIUM—*NB*—41

NITROGEN—*N*—7

NOBELIUM—*NO*—102

OGANESSON—*OG*—118

OSMIUM—*OS*—76

OXYGEN—*O*—8

PALLADIUM—*PD*—46

PHOSPHORUS—*P*—15

PLATINUM—*PT*—78

PLUTONIUM—*PU*—94

POLONIUM—*PO*—84

POTASSIUM—*K*—19

PRASEODYMIUM—*PR*—59

PROMETHIUM—*PM*—61

PROTACTINIUM—*PA*—91

RADIUM—*RA*—88

RADON—*RN*—86

RHENIUM—*RE*—75

RHODIUM—*RH*—45

ROENTGENIUM—*RG*—111

RUBIDIUM—*RB*—37

RUTHENIUM—*RU*—44

RUTHERFORDIUM—*RF*—104

SAMARIUM—*SM*—62

SCANDIUM—*SC*—21

SEABORGIUM—*SG*—106

SELENIUM—*SE*—34

SILICON—*SI*—14

SILVER—*AG*—47

SODIUM—*NA*—11

STRONTIUM—*SR*—38

SULFUR (Xanthous)—*S*—16

TANTALUM—*TA*—73

TECHNETIUM—*TC*—43

TELLURIUM—*TE*—52

TENNESSINE—*TS*—117

TERBIUM—*TB*—65

THALLIUM—*TI*—81

THORIUM—*TH*—90

THULIUM—*TH*—69

TIN—*SN*—50

TITANIUM—*TI*—22

TUNGSTEN—*W* (WOLFRAM)—74

URANIUM—*U*—92

VANADIUM—*V*—23

XENON—*XE*—54

YTTERBIUM—*YB*—70

YTTRIUM—*Y*—39

ZINC—*ZN*—30

ZIRCONIUM—*ZR*—40

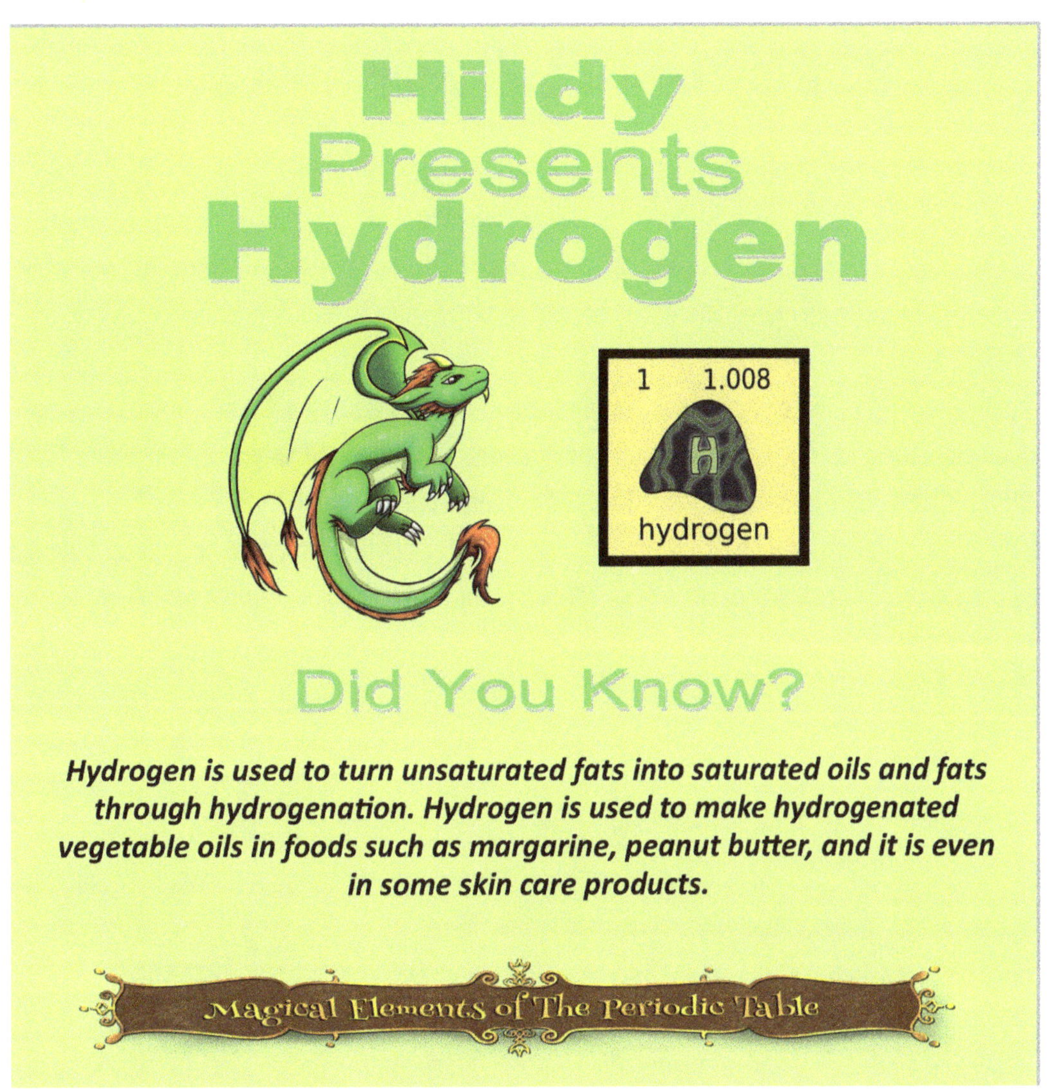

- Henry Cavendish studied hydrogen gas from 1766 to 1781. He was the first person to discover that when hydrogen burns, it makes water. This was an important scientific finding. Because of this, hydrogen got its name from the Greek word "hydrogen," which means "water-former."

- Hydrogen is often given colors to show how it is made. There is green, blue, grey, black, brown, pink, turquoise, yellow, and white hydrogen. Green hydrogen is seen as the cleanest because it is made using renewable energy, like wind or solar power. Black hydrogen is the most harmful to the environment because it is made from coal and creates a lot of pollution. The other colors show different ways hydrogen can be produced, and each one has a different level of impact on the planet. These color names help people understand which types are cleaner and which are dirtier.

- Hydrogen is different from the noble gases because it has only one electron in its outer shell. This makes it less stable and more ready to react. Noble gases already have full outer shells, so they do not usually join with other substances. When hydrogen gas is activated by heat, light, or other energy, it can react very quickly and give off heat. Hydrogen can also combine with oxygen and some other active gases. When it reacts with oxygen, it forms water vapor, which is harmless.

- Hydrogen, once called "inflammable air," became a key part of the great chemical changes of the late 1700s. Studying it helped scientists break an old idea: that water was a basic element. This pushed chemistry toward a new way of thinking, based on careful measurements and facts.

Did You Know? (continued)

- In 1783, hydrogen played a big part in a historic event. Once people learned how to gather it and make a lot of it, they used it in the first balloon flight with people onboard. This was an important moment because it helped show that humans could rise into the sky. It also proved that science was not just for study, but could create real change in everyday life. The flight helped start new ideas about travel, discovery, and what people could do with science.

- Seawater-to-hydrogen technology is becoming an important new way to cut pollution from ships. It works by using filtered seawater and renewable power to make hydrogen fuel right where it is needed. This means ships could stop using dirty bunker fuel and instead run on a cleaner fuel that gives off only water vapor when used. Because shipping creates a lot of harmful emissions, this idea could help reduce damage to the climate much sooner. It offers a practical step toward cleaner ocean travel and a lower-carbon future for the shipping industry.

- Science fiction is often the source of actual science. Many science fiction movies use hydrogen as a fuel or a key part of the story. In *Glass Onion: A Knives Out Mystery* (2022), the made-up fuel "Klear" is a solid form of hydrogen. In *Oblivion* (2013), machines called "hydro rigs" turn seawater into fusion power. In *The Martian* (2015), Mark Watney makes hydrogen from a chemical called hydrazine and uses it to make water. *The Hydrogen Age* (2004) is a film about hydrogen as a future energy source. *Quatermass 2* (1957) also has secret plants making strange materials.

- Hydrogen has been tied to a number of serious disasters over the years. Some happened in the sky, some in factories, and some during military tests. One of the best-known cases was the 1937 Hindenburg fire, when the German airship caught fire and 36 people died. Later, hydrogen explosions in industry also caused deaths and injuries. In 2019, an explosion at a hydrogen storage site in South Korea killed two people and hurt six others. That same year, a blast at a plant in Illinois killed four workers. In 2007, a hydrogen leak at a power plant in Ohio killed one worker and injured ten more. There were also dangerous accidents with hydrogen bombs. In Spain in 1966, a bomber carrying four bombs crashed, spreading radioactive material. In 1961, another plane broke apart, and one bomb almost went off. These events show how risky hydrogen can be when it is stored or handled badly.

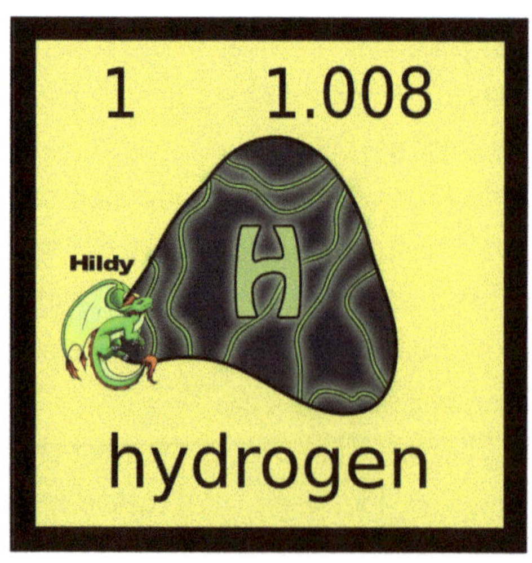

Structure Of Elements In The Periodic Table

Periodic tables are laid out in rows and columns.

Each element is placed in a specific location because of its atomic structure. Elements are arranged in Families.

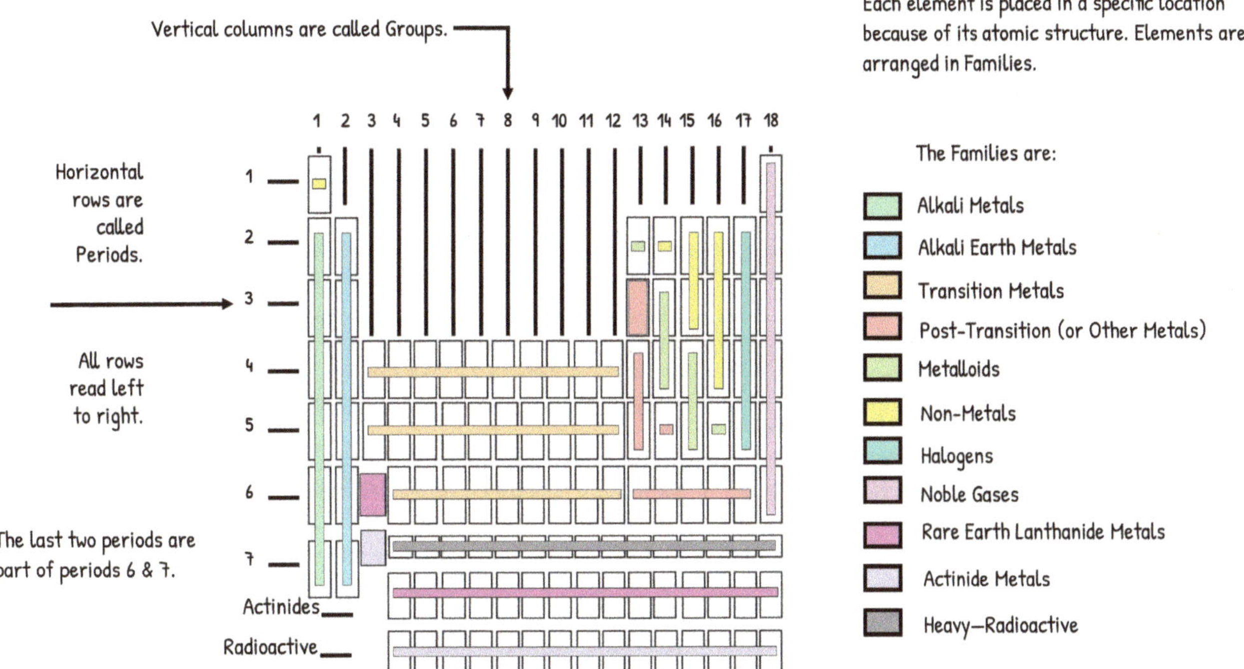

The term 'Element' is used to describe atoms with specific characteristics.
Every element in the first column or Group has 1 electron in the outer orbital (shell).
Every element in the second column (group two) has two electrons in the element's outer orbital.
The number designation of each Group represents the number of electrons in the element's outer orbital—
except for Group 18, Period 1—Helium. It only has 2 electrons.
Those electrons, called Valence Electrons, are what chemically bond with other elements.

Atomic Structure of Element: The atomic structure of an element refers to the arrangement of protons and neutrons in the nucleus of the atom, and the electrons in the electron cloud around the nucleus. Group 1, Period 1—Hydrogen is the only element that has no neutrons.

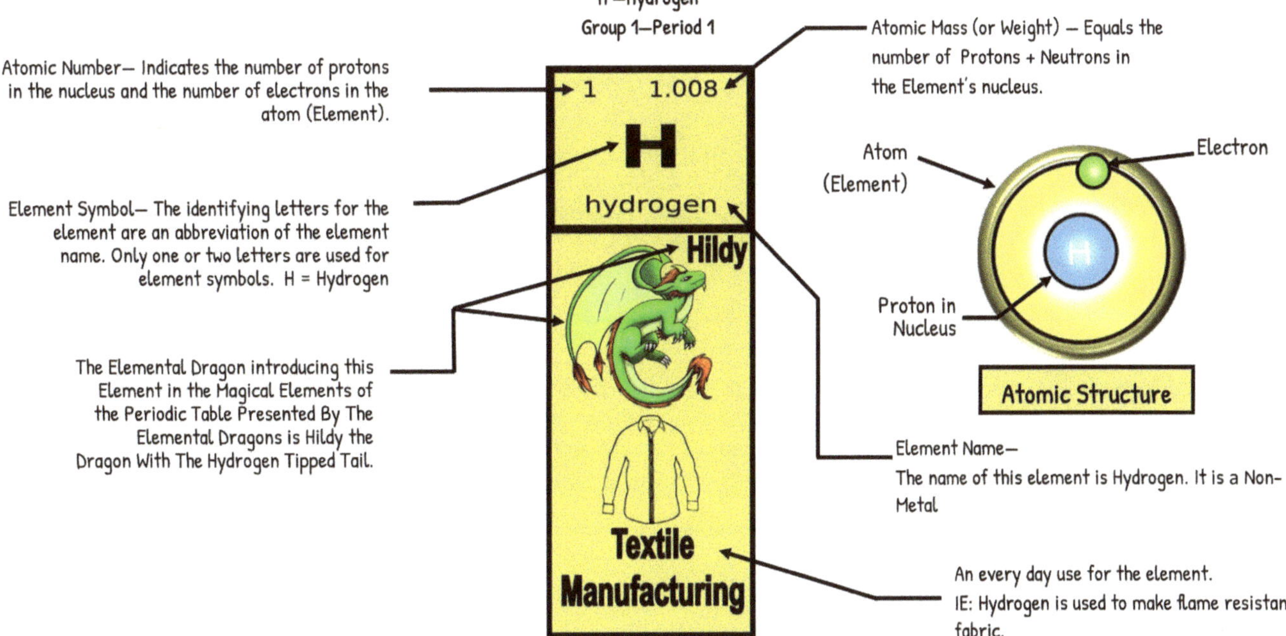

Types of Elements On The Periodic Table

Alkali Metals—Some metals on the periodic table are soft and shiny. They are so soft that they can be cut with a knife! These metals are excited to give away electrons to elements in need, making them highly reactive. Ther electron transfer creates a compound known as a salt. Surprisingly, these metals are not found in nature alone; they must be extracted from other sources. Examples of these metals include lithium, sodium, potassium, rubidium, cesium, and francium.

Alkali Earth Metals—The elements in column 2 of the periodic table have 2 outer electrons in their shell. Ther makes them very active with nonmetals that need electrons to stay stable. When they react, they make something called a salt. They are often found in nature all by themselves, and they can even conduct electricity. The elements are beryllium, magnesium, calcium, strontium, barium, and radium.

Post-Transition (or other Metals)— Elements directly to the right of the transition metals. They are known as "poor metals: and are soft and brittle. These include aluminum, gallium, indium, tin, thallium, lead, bismuth, zinc, cadmium and mercury.

Transition Metal—The main metals are found in the middle and bottom rows of the periodic table. They look like metal, can conduct electricity, can bend and be shaped easily. The period 4 transition metals are scandium, titanium, vanadium, chromium, manganese, iron, cobalt, nickel, copper, and zinc. The period 5 transition metals are yttrium, zirconium, niobium, molybdenum, technetium, ruthenium, rhodium, palladium, silver, and cadmium. The period 6 transition metals are lanthanum, hafnium, tantalum, tungsten, rhenium, osmium, iridium, platinum, gold, and mercury. The period 7 transition metals are the naturally-occurring actinium, and the artificially produced elements rutherfordium, dubnium, seaborgium, bohrium, hassium, meitnerium, darmstadtium, and roentgenium.

Metalloids—The elements called metalloids are a mix of metals and nonmetals. They look like metals, but can't conduct electricity very well. They also break easily and act like nonmetals. These include boron, silicon, germanium, arsenic, antimony, tellurium, astatine, and polonium.

Non-Metals—These elements reside in columns 15-17, and can be gases, liquids, or solids. They don't conduct heat or electricity. The solids are brittle, and they have no metallic luster. They readily accept electrons from metals to form salts. These include nitrogen, oxygen, fluorine, chlorine, bromine, and iodine.

Halogens—Halogen chemicals are a special type of element. When they mix with metal, they become a kind of salt. Halogens are super reactive because they like to take an electron from metals. They can be found in column 17 of the element table. Some of them can be found in nature, but most are very dangerous and can hurt you if you touch them. They include fluorine, chlorine, bromine, iodine, and the radioactive elements astatine and tennessine.

Noble Gases—These elements reside in column 8. They are all odorless, colorless gases that are chemically very stable (inert). They don't generally form compounds by bonding with another element. These include helium, neon, argon, krypton, xenon, and radon.

Lanthanide Rare Earth Minerals—The Japanese call them "the seeds of technology." The US Department of Energy calls them "technology metals." These elements have atomic numbers 57-71. They are vital to industry. They can be added to metals to strengthen them to make alloys such as stainless steel, used to refine crude oil, and are crucial in producing technology—electronics, telecommunications, and metal devices to name a few. They are lanthanum, cerium, praseodymium, neodymium, promethium, samarium, europium, gadolinium, terbium, dysprosium, holmium, erbium, thulium,

Actinide Metals—Any of a series of chemically similar metallic elements with atomic numbers ranging from 89 (actinium) to 103 (lawrencium). All of these elements are radioactive, and two of the elements, uranium and plutonium, are used to generate nuclear energy. The lanthanides and actinides are sometimes called the inner transition metals, referring to their properties and position on the table. They are actinium, thorium, protactinium, uranium, neptunium, plutonium,

Super Heavy—Radioactive—Superheavy elements are those elements with a large number of protons in their nucleus. Elements with more than 92 protons are unstable; they decay to lighter nuclei with a characteristic half-life. They do not occur in large quantities (if at all) naturally on earth, and only exist briefly under highly controlled circumstances. They include lawrencium, rutherfordium, dubnium, seaborgium, bohrium, hassium, meitnerium, darmstadtium, roentgenium, copernicium, nihonium, flerovium, moscovium, livermorium, tennessine, and oganesson.

Alloys

An **alloy** is a mixture of chemical elements of which at least one is a metal. An alloy is a solid. Unlike chemical compounds with metallic bases, an alloy will retain all the properties of a metal in the resulting material, such as electrical conductivity, ductility, opacity, and luster, but may have properties that differ from those of the pure metals, such as increased strength or hardness. In some cases, an alloy may reduce the overall cost of the material while preserving important properties. In other cases, the mixture imparts synergistic properties to the constituent metal elements such as corrosion resistance or mechanical strength. Some of the most common alloys are

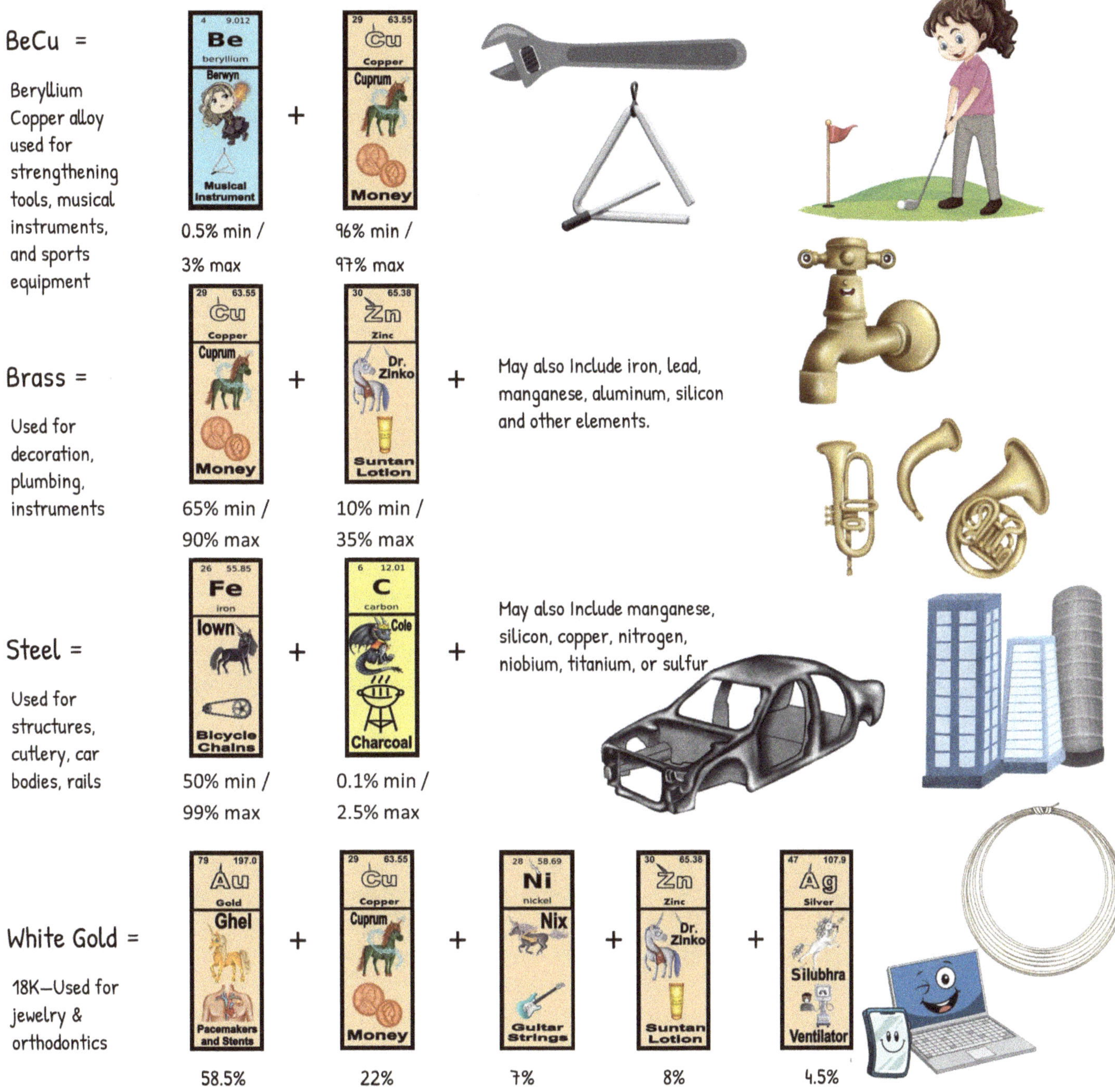

BeCu =

Beryllium Copper alloy used for strengthening tools, musical instruments, and sports equipment

Brass =

Used for decoration, plumbing, instruments

Steel =

Used for structures, cutlery, car bodies, rails

White Gold =

18K—Used for jewelry & orthodontics

Some other common alloys are Bronze, Cast Iron, Cupronickel, Magnalium, Mischmetal, Nichrome, Nitinol, Pewter, Solder, Sterling Silver and Tungsten Carbide.

The above chart only shows a few of the hundreds of metal combinations. For instance, 24 carat gold is a pure naturally occurring yellow metal. There are four basic shades of gold alloys: yellow gold, white gold, rose gold, and green gold. A huge range of other colored golds are also possible, including red (gold and copper), grey (gold, iron and copper), purple (gold and aluminum), blue (gold and iron) and black (gold and cobalt), depending on the amounts of different metals alloyed together.

Compounds

A **compound** is a substance formed due to the chemical union (a chemical reaction) between two or more atoms or molecules. Ideally there should be two or more elements to form a compound. Most of the compounds are from a non-metallic origin.

Fluorocarbon =

Used for— waterproofing agents, lubricants, sealants, and leather conditioners.

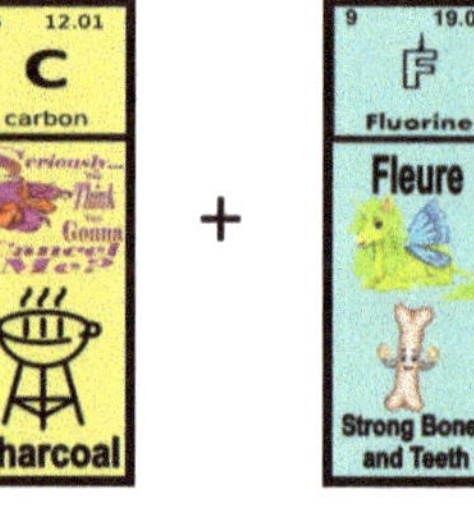

Both Carbon and Fluorine are reactive nonmetals.

Sodium Fluoride =

Used for—the fluoridation of drinking water, in toothpaste, in metallurgy, and as a flux, and is also used in pesticides and rat poison.

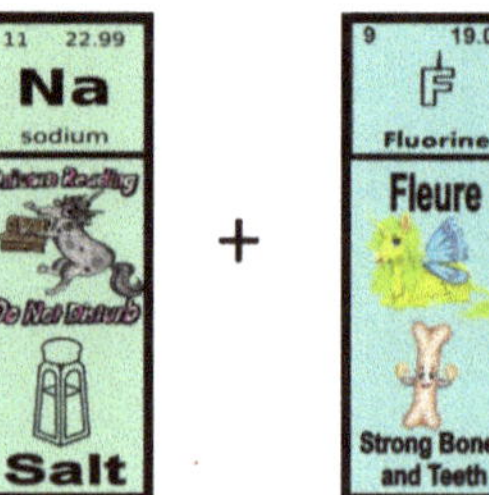

Sodium Fluoride is a simple ionic compound, made of the sodium (Na+) cation and fluoride (F-), an anion of Flourine.

Potassium Iodide =

It's medical use is to block absorption of radioactive iodine by the thyroid gland.

Potassium iodide (also called KI) is a salt of stable (not radioactive) iodine. Potassium is an alkali metal and Iodine is a reactive nonmetal.

Calcium Bromate =

It is used as a bread dough and flour "improver" or conditioner

Calcium is an alkaline earth metal and Bromine is a reactive nonmetal.

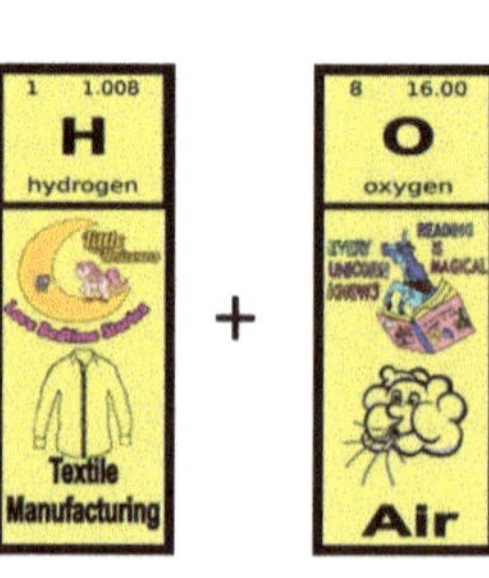

= ??????

(Answer can be found below.)

Can you guess the most important compound of all?

The above chart only shows a few of the millions of combinations of elements that create compounds.

Iodine, alone has 37 compounds. Some of them are Ammonium iodide (NH4I), Cesium iodide (CsI), Copper(I) iodide (CuI), Hydroiodic acid (HI), Iodic acid (HIO3), Iodine cyanide (ICN), Iodine heptafluoride (IF7), Iodine pentafluoride (IF5), Lead(II) iodide (PbI2), Lithium iodide (LiI), Nitrogen triiodide (NI3), Potassium iodate (KIO3), Potassium iodide (KI), Sodium iodate (NaIO3), and Sodium iodide (NaI).

Hydrogen is used in more compounds (nearly 100) than any other element because it can form bonds with almost all metals, metalloids, and non-metals. Some of the most common are water (H2O), Hydrogen Peroxide (H2O2) and table sugar (C12H22O11).

Neon forms no known compounds.

Definitions

Atomic Structure of Element: The atomic structure of an element refers to the arrangement of protons and neutrons in the nucleus of the atom, and the electrons in the electron cloud around the nucleus.

Atomic Number: An element's atomic number refers to the number of protons it has in its nucleus. In a neutral atom the number of protons always equals the number of electrons.

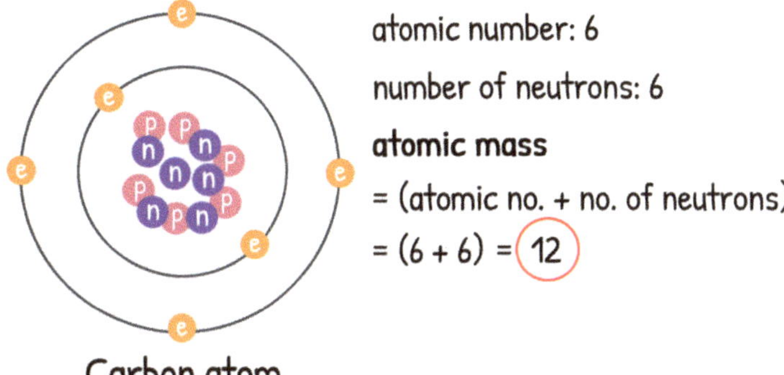

Atomic Weight (Mass) of Element: The atomic mass of an element is how heavy it is. It is made up of protons and neutrons that are in the middle of the element. Some elements have different versions with different amounts of neutrons, but they still have the same amount of protons. The atomic mass is the average weight of all these versions of the element.

Allotrope: Allotropes are different forms of an element that look and act different, but are made of the same stuff. Some elements have more than one form. For instance, carbon can be a shiny diamond or a gray pencil lead called graphite.

Isotope: Isotopes are different types of atoms that have the same parts, like protons and electrons, but they have a different number of neutrons. For example, the three most stable isotopes of hydrogen: protium (A = 1), deuterium (A = 2), and tritium (A = 3).

Crystalline Structure of Element: The crystalline structure of an element is how its atoms, ions, or molecules stick together in a pattern to make a cool crystal shape.

Ferrous and Non-Ferrous Metals: When we say ferrous metal, it means that iron is a big part of the metal. But if there's only a little bit of iron in the metal, we call it non-ferrous. The word "ferrous" comes from Latin and means iron, which is why iron's symbol is Fe.

Ductile Metals: These are capable of being made into long, thin wire or thread. Copper and Silver are ductile metals.

Malleable Metals: These can be hammered or rolled into thin sheets without cracking or breaking. Gold is malleable.

Ferromagnetic: Materials that are strongly attracted to a magnet. Such materials can be permanently magnetized. These include the elements iron, nickel and cobalt and their alloys, some alloys of rare-earth metals, and some naturally occurring minerals such as lodestone.

Magnetostriction—Ther is the term for a special thing that happens to magnetic materials. When these materials get turned into magnets, they also change their shape or size.

Paramagnetic: Slightly attracted to a magnetic field, but do not retain magnetic properties once the field is removed.

Diamagnetic: Slightly repelled by a magnetic field, but do not retain magnetic properties once the field is removed.

Electrical Properties: Conductor—a thing that lets electricity flow through it. Semi-conductor—a special material (usually silicon) that can conduct electricity, but not as well as metal. Insulator (non-conductor)—a material (usually glass) that stops electricity from flowing.

Reactive Gas: These gases are really good at reacting with stuff! They are called "sticky gases" because they can react to things like plastic and wet surfaces when they touch them. These are nitrogen, oxygen, hydrogen, carbon dioxide, fluorine, and chlorine.

Non-Reactive Gas: An inert gas is like a super shy gas that doesn't like to hang out with other chemicals. It doesn't make any new friends by reacting with them, so it doesn't form any chemical compounds. We also call these special gases "noble gases."

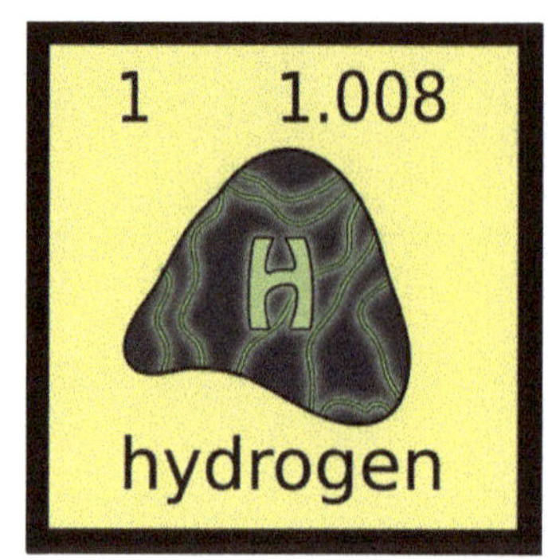

What Makes Hydrogen Seem Magical?

The Periodic Table is a treasure trove of knowledge, containing all the building blocks that make up our universe. Among its diverse elements, hydrogen holds a special place. This humble atom, with its simple structure of a single proton and electron, possesses extraordinary properties that make it seem almost magical in nature. From its role in fueling stars to its potential as a clean energy source, hydrogen continues to captivate scientists and inspire wonder.

One of the most remarkable aspects of hydrogen lies in its ability to power the stars. In the heart of these celestial giants, like our sun, hydrogen atoms fuse together under immense pressure and temperature, releasing a staggering amount of energy in the process. This nuclear fusion reaction, known as the proton-proton chain, provides the star with its life-sustaining heat and light. Hydrogen's pivotal role in this cosmic dance of creation and destruction captures our imagination and emphasizes its truly magical nature.

Moreover, hydrogen possesses the exceptional ability to exist in three different states simultaneously - gas, liquid, and solid - depending on temperature and pressure. The gaseous form, commonly observed on Earth, is so light that it rises rapidly, making it a vital component in rocket fuel. Liquid hydrogen, on the other hand, is incredibly cold, boiling at a temperature of -423.17°F (-252.87°C). Due to its remarkable combination of density and temperature, liquid hydrogen has been crucial in propelling space shuttles into orbit, further highlighting its enchanting properties.

Another captivating feature of hydrogen is its potential as a clean and sustainable energy source. Hydrogen fuel cell technology has the ability to revolutionize our transportation systems by producing electricity with water vapor as the only byproduct, eliminating harmful emissions. This clean energy alternative offers a glimmer of hope for combating climate change and reducing our dependence on fossil fuels.

Hydrogen's fascinating chemical properties give it the ability to easily combine with other elements, forming a diverse range of compounds. It can react with oxygen to create water, a vital component for life on Earth. It can also combine with carbon to form hydrocarbons, the fundamental building blocks of organic compounds. These versatile chemical interactions not only illustrate the element's versatility but also demonstrate its magical ability to participate in the creation of diverse and complex substances.

Hydrogen's cosmic presence extends beyond Earth. This extraordinary element plays a significant role in the vastness of our universe, being a primary component of interstellar clouds and actively participating in the formation of new stars and planetary systems. This celestial connection adds yet another layer of mystique to the element, fueling our curiosity about the cosmos.

Meet Hildy, The Dragon with a Magical Tail Tipped with Hydrogen

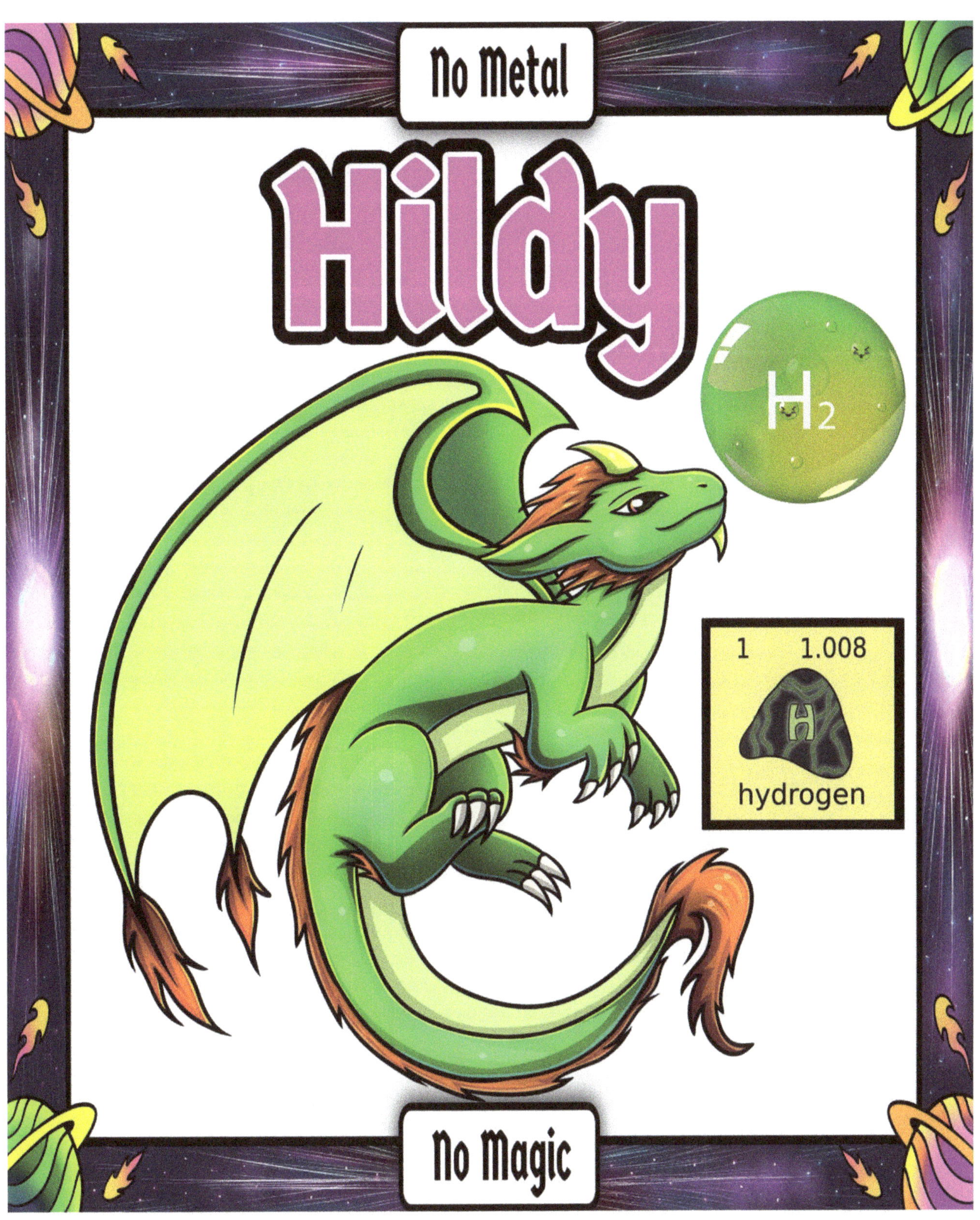

Hildy the elemental dragon woke to a MarBryn sunrise painted in copper and jade, her consciousness blooming like a flower unfurling toward warmth. Her skin gleamed a vibrant green, a friendly brightness that hinted at mischief and warmth. A pale lime underbelly matched the light-green membrane of her enormous wings, which stretched like sails above a coiled, orange-furred tail that twitched with anticipation. Her orange mane sparked with energy, crackling softly as morning dew evaporated against her scales, and a playful glint lit her amber eyes as she tested the air for adventure.

She carried a rare gift: the ability to manipulate hydrogen in all its knowing forms. Hydrogen wasn't just air to her—it was an ember that could be coaxed into shape, a mist she could condense, a fusion flame she could tame. She had spent centuries learning this art, studying the invisible dance of atoms, understanding the language that elements spoke to those patient enough to listen.

A quick-witted sprite named Brin flew near her shoulder in a swirl of emerald light, humming a tune that seemed to make the morning brighter. Brin's tiny wings buzzed with curiosity, and he spoke in spark-crackling riddles. Their bond was a study in balance: Hildy's power could ignite the night, while Brin's clever schemes kept them witty, humble, and prepared. They had been companions for three seasons now, ever since Brin had appeared in her cave during a storm, soaked and shivering, speaking in riddles even as his teeth chattered.

"Good morning, captain," Brin chirped, his voice like wind chimes in a gentle breeze. "I dreamed of storms last night. Dark ones. The kind that don't belong."

Hildy's expression grew serious. She had sensed it too—a disturbance in the hydrogen flows, a wrongness in the air currents. "The Smokewraith Caverns," she murmured.

Their enemy emerged from the Smokewraith Caverns that very afternoon: Malvar, a powerful apprentice to the most evil sorcerer, Magh, the scourge of MarBryn. Malvar believed hydrogen belonged to the dark arts and should be hoarded by those with the will to dominate. He wove dangerous currents of gas with twisted intent, threatening towns with fizzing storms and blistering heat that left scorch marks on the earth. Malvar sought to weaponize hydrogen fusion to become a master of combustion, to reshape the world according to his vision of power and control. He had already destroyed three villages on the eastern plains, leaving behind only ash and the acrid smell of his ambition.

The first rumor spoke of a flame that wouldn't die, a fire that consumed everything in its path. Hildy leaped from cliff to cliff, her powerful wings propelling her across the landscape, summoning and dispersing hydrogen with precision. She condensed invisible molecules, their droplets swirling around her like a halo, and whispered to them in the ancient language of elements. With a breath, she gathered hydrogen, liquefying it mid-air into a shimmering pool that powered a portable forge Brin could use to craft gear for their journey—lighter armor, stronger tools, better maps marked with safe routes.

She demonstrated hydrogen levitation, lifting herself above jagged rocks to scout ahead, Brin riding the thermals with excited chatter. From high above, they could see the path of destruction Malvar had carved through the land: blackened forests, rivers boiled dry, stone turned to glass by the intensity of his flames.

"He's heading toward Millbrook," Brin observed, pointing with one tiny finger. "The village where the children gather for the harvest festival."

Hildy's jaw tightened. "Then we fly faster."

As they gained altitude, Brin nudged her shoulder. "Race you to the ridge, Hildy! I'll keep the map, you keep the wind in your sails."

"Always," she teased, unfurling her wings with a sound like silk sliding across stone. With a surge of hydrogen, she pressed off the air, lifting into a graceful arc. The valley fell away below as she glided, the wind singing in her ears, the morning light catching the iridescence of her scales. Brin clapped his hands in spark-crackles, "Look at you go, captain of the currents! No one moves like you do!"

They arrived at Millbrook just as smoke began to rise from the northern edge of town. Malvar had already begun his assault, sending waves of superheated gas toward the wooden buildings. But Hildy was faster. She descended like a comet, her wings spread wide, and positioned herself between the sorcerer and the village.

"Enough, Malvar," she called, her voice carrying the weight of her years and her purpose. "Hydrogen is not your weapon to command."

Malvar laughed, a sound like breaking glass. "The little dragon comes to play. How delightful. And you've brought a pet sprite. How... quaint."

At the heart of the mountains—where Malvar had retreated to make his final stand—Hildy faced him in a cavern vast as a cathedral. He unleashed torrents of hydrogen gas, hoping to suffocate them in a cathedral of steam and ash, to overwhelm her with sheer volume and heat. Hildy answered with a controlled explosion, shaping the heat into a radiant shield that kept Brin safe behind a barrier of compressed, cool air. She summoned a precise fusion spark, coaxing two hydrogen atoms to merge with surgical precision, releasing a clean, contained pulse of energy that destabilized Malvar's storm and sent him sprawling backward.

Brin darted in, tossing a compressor of crystallized energy dust that disrupted Malvar's control over the gas around him. Hildy's pinkish-orange mane flared as she extended her tail, the flame-tipped tip catching sparks that suited the moment perfectly. She drew the gas into a focused jet, extinguishing Malvar's flames before they could claim the valley below. The crowd—villagers who had followed at a safe distance—exhaled as the battlefield cleared and the storm paused at the horizon, dissipating like a bad dream.

Malvar collapsed, his power spent, his will broken. He lay gasping on the stone floor, finally understanding that raw power without wisdom was merely destruction waiting to happen.

"Nice work, Brin," Hildy said, landing beside a rock face that glowed softly with residual hydrogen heat, her breathing steady despite the intensity of the battle. "Your riddles saved the day again."

Brin grinned, doing a loop in the air. "Only because you give me air to think with. And fuel to keep up. Besides, I told you that riddle about the dragon and the sorcerer last week. Turns out it was prophetic."

A sudden gust whipped through the canyon, and Malvar's voice hissed from the shadows where he lay defeated. "You can't outrun destiny, dragon. Hydrogen obeys the biggest plan."

Hildy raised her head and spoke with quiet resolve, her amber eyes glowing softly. "Hydrogen isn't a weapon to wield blindly. It's a tool to heal, to light a path forward, to bring people home. That's the plan that matters."

In the aftermath, as the sun began to set, Brin hovered close as Hildy practiced her hydrogen healing, sending a warm wave of energy through a wounded villager's shoulder. The sprite's eyes softened as he watched the color return to the villager's face, the pain fading from their expression. "You're a force for renewal, Hildy. Not just power. You are hope in flight."

The valley settled into peace, and Brin sang songs of gratitude in his crackling voice throughout the evening. Hydrogen had become a beacon of renewal, and Hildy's legend grew as a dragon of warmth, intelligence, and steadfast courage. The villagers began to understand that the greatest strength wasn't in domination, but in stewardship.

"Ready for the next flight, partner?" Hildy asked the following dawn, flexing her wings in the light as the sun painted the sky in gold and amber.

Brin nodded, eyes sparkling with anticipation and joy. "Always. The sky is wide, and the hydrogen flows like a song. Let's ride it to the far horizon and see what wonders await."

Hildy crouched, her muscles coiling like springs, then launched into the clear morning air with a powerful thrust of her wings. They rose as one, the wind threading through her green scales, Brin's laughter trailing behind like sparks dancing in their wake. "To the next adventure!" Brin cried, his voice carrying on the wind.

Hildy answered with a confident smile, her hydrogen-tipped tail curling behind her like a living flame, graceful and purposeful. "Where hydrogen leads, we follow." And they vanished into the bright, misty sunrise, ready for whatever lay beyond the mist, their bond unbreakable and their purpose clear.

Enjoy This Coloring Page Featuring

Hildy The Dragon with the Hydrogen Tipped Tail

Sample Page From Magical Elements of the Periodic Table

Presented By The Elemental Dragons Book

Hildy Presents Hydrogen

Symbol: H Atomic Number: 1 Atomic Mass: 1.008

Hydrogen Facts

- First artificially produced in the early 16th century
- The lightest gas
- Poor conductor of heat and electricity
- Very weakly magnetic—Diamagnetic
- H2 (molecular Hydrogen) is highly reactive
- Non-Metal Gas

Green Hydrogen is produced by using electricity to split water molecules (H2O) into hydrogen (H2) and oxygen (O2). Blue hydrogen is produced from hydrocarbon sources, such as natural gas, and combined with carbon capture technology.

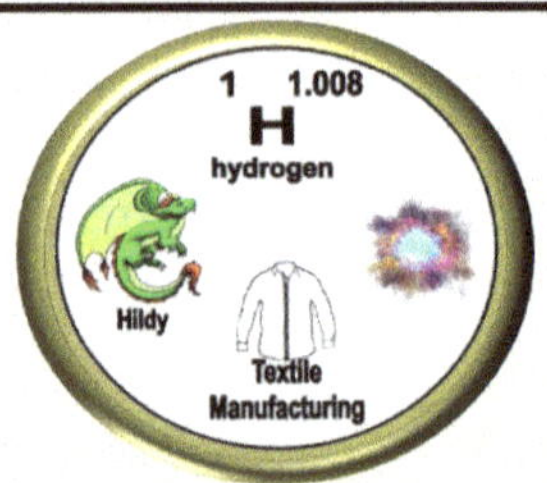

Hildy's Magical Abilities

- A rainbow blast of energy from Hildy's Hydrogen tipped tail provides super fuel to power vehicles from race cars to space craft.

Atomic Structure

Uses For Hydrogen

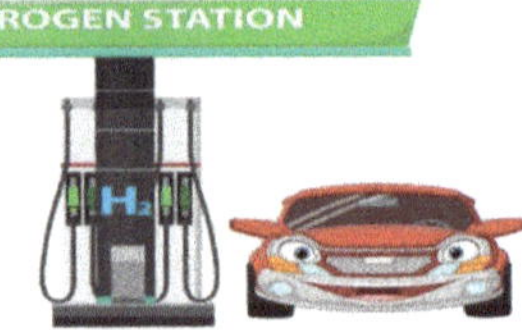

Hydrogen is used to fuel vehicles in some countries.

Hydrogen is used in the hydrocracking process which takes heavier refinery products like crude oil and "cracks" the large molecules into smaller ones to make things like gasoline, diesel and thousands of other chemicals.

Next to oil refineries, ammonia is currently the largest application of hydrogen. Ammonia (NH3) is used to produce ammonium nitrate, a fertilizer, and is part of many household cleaning products. Don't forget to wear your gloves when using it!

Hydrogen is used to turn unsaturated fats into saturated oils and fats through hydrogenation. Hydrogen is used to make hydrogenated vegetable oils in foods such as margarine, peanut butter, and it is even in some skin care products.

Did You Know?

- Henry Cavendish, during the years 1766-1781, was the first person to find out that hydrogen gas turns into water when it is burned. Because of this, it got its name from the Greek word "hydrogen" which means "water-former".
- Hydrogen is color coded. There's green, blue, grey, black and brown, pink, turquoise, yellow and white. Green is the cleanest and black is the most environmentally damaging.
- Hydrogen is not like the noble gases because it only has one electron in its shell, which makes it unstable. Unlike noble gases that have full outer shells of valence electrons, hydrogen is different. When hydrogen gas gets activated by heat, light, or other things, it reacts very fast and releases heat. It can also combine with oxygen and other reactive gases to form harmless water vapor.

MEET THE CLAN

THE DRAGONS WITH THE MAGICAL ELEMENTAL TIPPED TAILS

Magical Elements of The Periodic Table

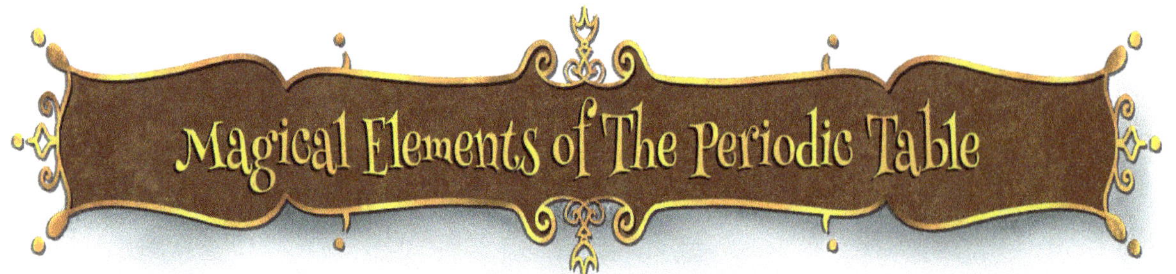

Create Your Own Magical Dragon Elemental

Hildy The Dragon With The Hydrogen Tipped Tail

Symbol: H Atomic Number: 1 Atomic Mass: 1.008

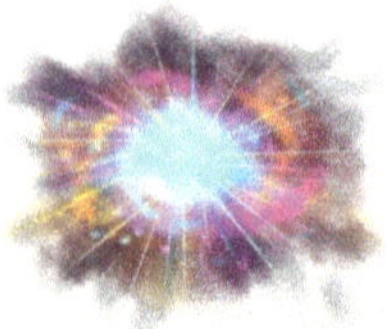

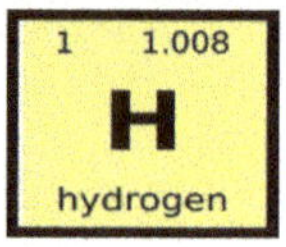

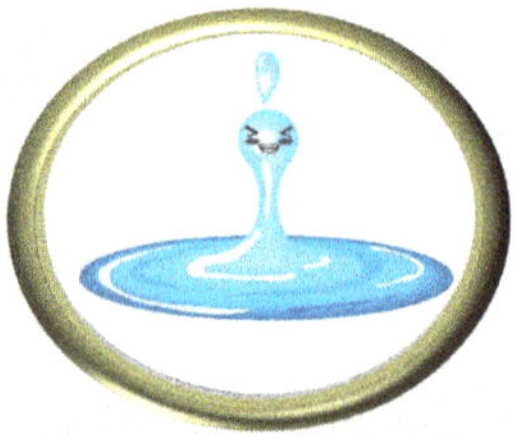

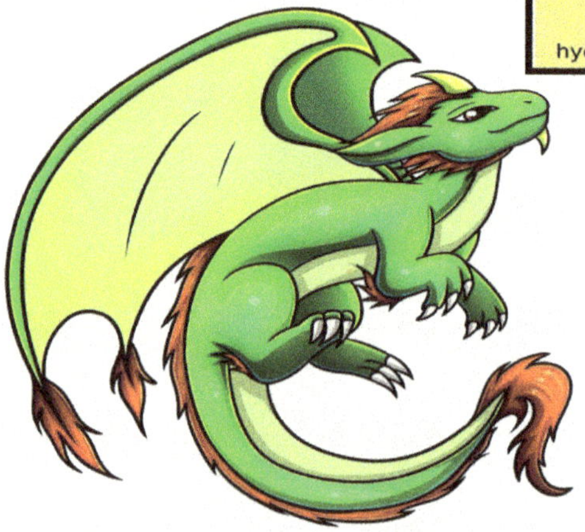

Magical Clan Crest Symbol

Occurs in water molecules

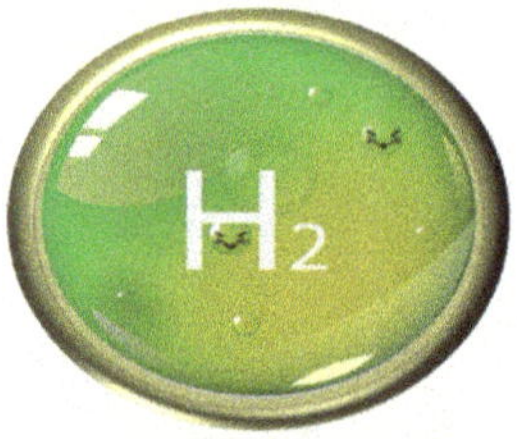

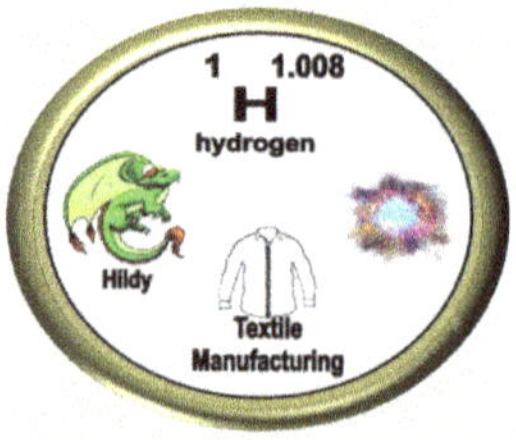

Non-Metal Gas

Hildy's Magical Abilities

- A rainbow blast of energy from Hildy's Hydrogen tipped tail provides super fuel to power vehicles from race cars to space craft.

Hydrogen Periodic Symbol

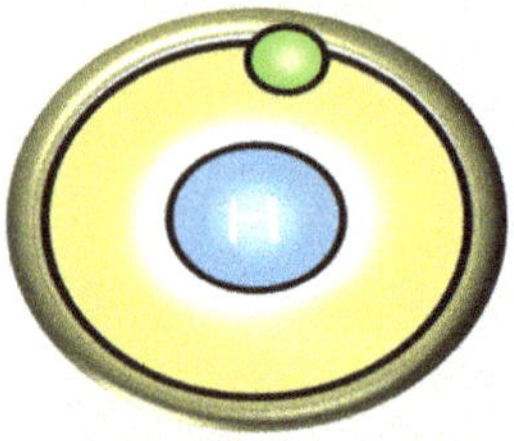

Atomic Structure

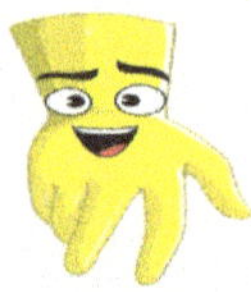

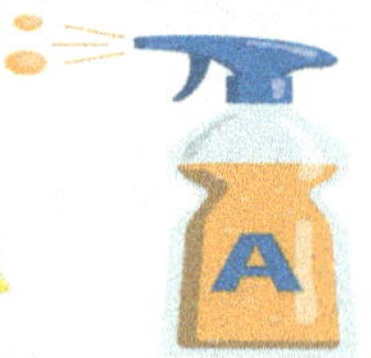

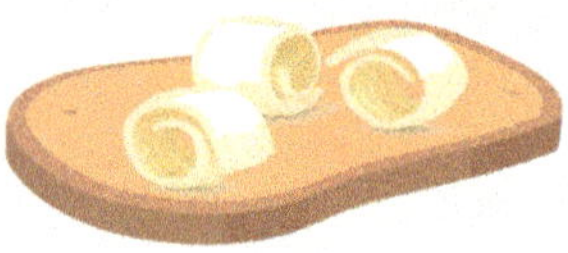

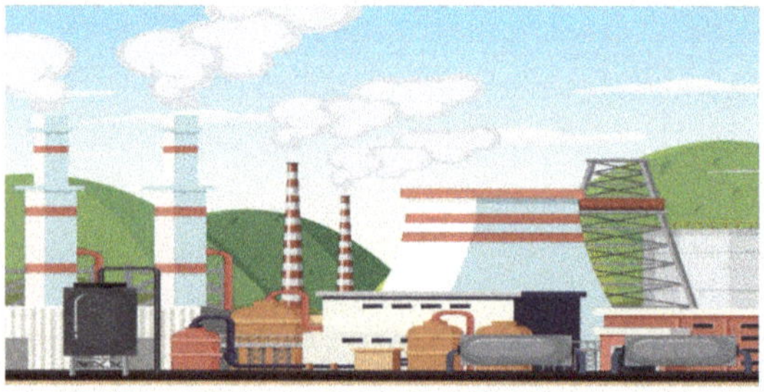

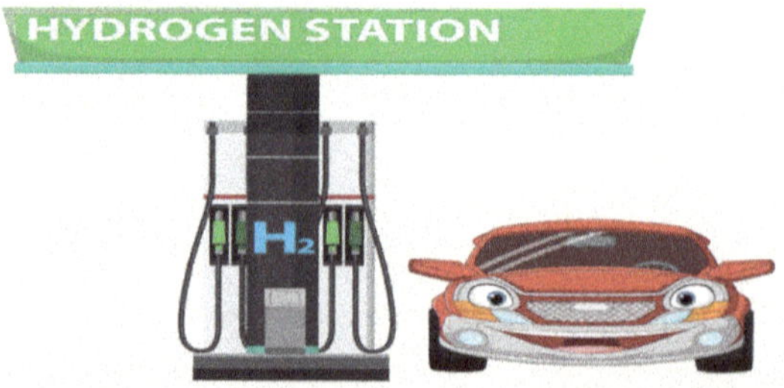

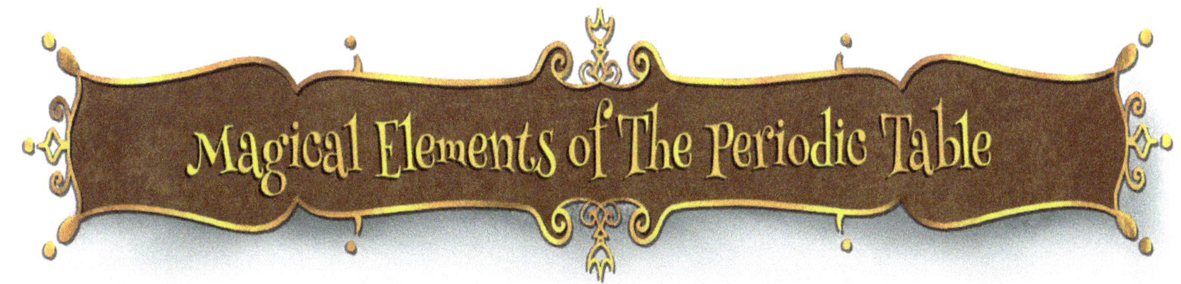

Students may either use a program like power point to cut and paste clip art into a Magical Dragon Elemental Blank or, if they wish, they may draw everything themselves.

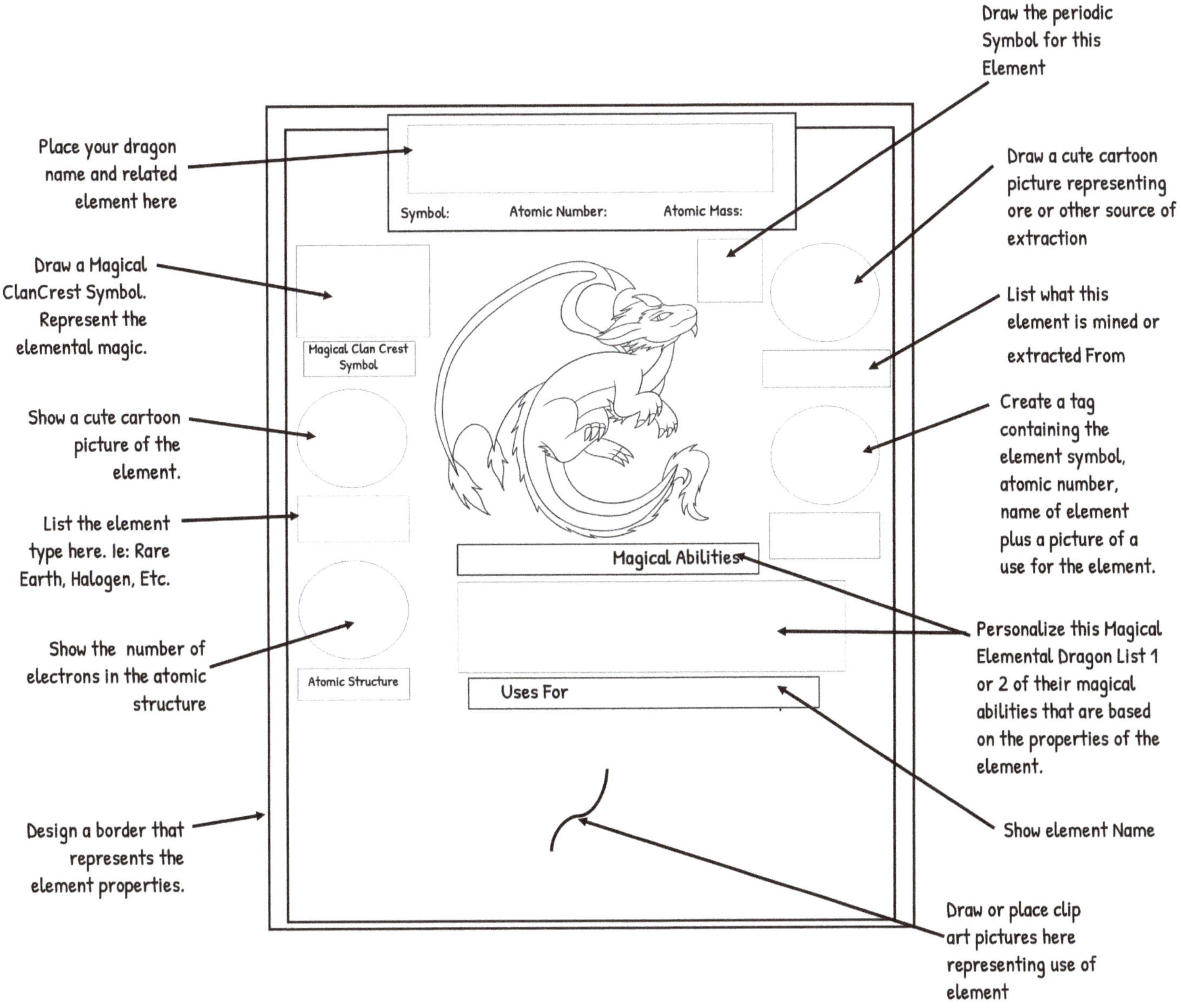

Place your dragon name and related element here

Draw a Magical ClanCrest Symbol. Represent the elemental magic.

Show a cute cartoon picture of the element.

List the element type here. Ie: Rare Earth, Halogen, Etc.

Show the number of electrons in the atomic structure

Design a border that represents the element properties.

Draw the periodic Symbol for this Element

Draw a cute cartoon picture representing ore or other source of extraction

List what this element is mined or extracted From

Create a tag containing the element symbol, atomic number, name of element plus a picture of a use for the element.

Personalize this Magical Elemental Dragon List 1 or 2 of their magical abilities that are based on the properties of the element.

Show element Name

Draw or place clip art pictures here representing use of element

Symbol: Atomic Number: Atomic Mass:

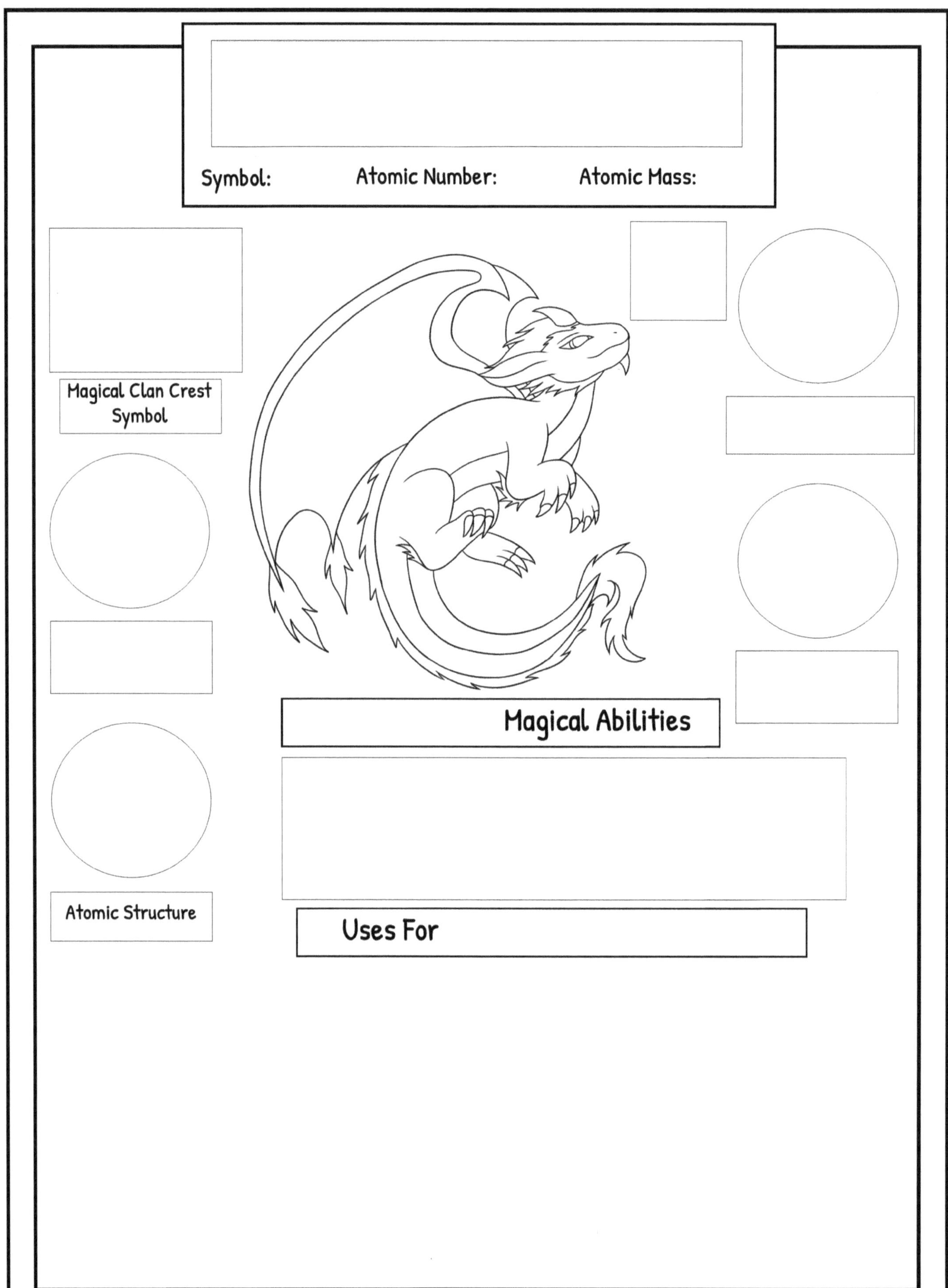

Magical Clan Crest
Symbol

Atomic Structure

Magical Abilities

Uses For

Magical Dragon Elemental Research Sheet

Before starting your Magical Dragon Elemental graphics page, do some research on your chosen element.

Name of Magical Dragon:	
Dragon's Magic Power Based on the Element's Properties:	
Magical Clan Crest Symbol:	
Element Name:	
Element Symbol:	
Atomic Number:	
Atomic Mass:	
What year and where was this Element discovered?	
Who discovered this Element?	
Element Group:	
Element Period:	
Element Family Name:	
State of Element At Room Temperature:	
What is Element Mined or Extracted From?	
Is Element Magnetic?	
Does Element Conduct Electricity?	
Where is the Element commonly found in Nature?	
What is 1 alloy of the Element? How used?	
What is 1 compound of the Element? How used?	
Name the most common use for this Element:	
Name a little known use for this Element:	
Name one more use for this Element:	
Interesting and Fun Facts:	

Magical Unicorn Elemental Research Sheet

Before starting your Magical Unicorn Elemental graphics page, do some research on your chosen element.

Name of Magical Unicorn:	Ghel The Gold Horn Unicorn
Unicorn's Magic Power Based on the Element's Properties:	Ghel can see past, present and future. She is empathic and can sympathize with the feelings of other. They say she has a heart of gold.
Magical Herd Crest Symbol:	An open heart with a Celtic Trinity Knot.
Element Name:	Gold
Element Symbol:	Au— Comes from Aurum which is the Latin word for Gold.
Atomic Number:	79
Atomic Mass:	196.97
What year and where was this Element discovered?	Around 4,600 BCE in Bulgaria
Who discovered this Element?	Unknown
Element Group:	11 on Periodic TAble
Element Period:	6 on Periodic TAble
Element Family Name:	Gold is a Noble Transition Metal
State of Element At Room Temperature:	Solid
What is Element Mined or Extracted From?	Quartz Veins. It is also found in gravel in streams.
Is Element Magnetic?	It is Diamagnetic. It's only weakly magnetized when placed in a magnetic field.
Does Element Conduct Electricity?	Gold is a great electrical conductor used in printed circuitry of computers.
Where is the Element commonly found in Nature?	One of the largest deposits is found in the United States in Arkansas.
What is 1 alloy of the Element? How used?	White gold is an alloy of gold, palladium, nickel and zinc.
What is 1 compound of the Element? How used?	Gold Phosphide is a semiconductor used in high power, high frequency applications and in laser diodes.
Name the most common use for this Element:	Jewelry
Name a little known use for this Element:	Acupuncture needles
Name one more use for this Element:	Gold is used in airbags in cars.
Interesting and Fun Facts:	Gold was used in ancient Egypt to fill decayed teeth. Gold thread is incorporated in astronaut spacesuits to protect them from the heat of the sun.

Write a paragraph below to describe your magical dragon elemental. Based on the information obtained from research of your chosen element, how did you determine your dragon's name? What are your dragon's magic powers? What are their likes/dislikes, strengths/weaknesses, personality traits? What color is your dragon and why did you pick that color? What is your dragon's Clan Crest Symbol?

Magical Unicorn Elemental Sample Description

Ghel The Gold-Horned Unicorn

This magical unicorn has a golden horn and hooves that glow like the sun. Her hide is honey-gold and her flowing mane and tail are golden-blonde.

Ghel is a member of the Metal Horn Unicorn Tribe from Unimaise. Gold is linked to the heart chakra because it holds a warm energy that brings soothing vibrations to the body to aid in the healing process. Ghel's Magical Herd Crest symbol is an open heart with a Celtic Trinity knot which indicates that she can see past, present and future. She is empathic and can sympathize with the feelings of others.

Being empathic does not make Ghel a weakling. She is brave with strong opinions and is a true champion to those she loves. She is known as the unicorn with the "heart of gold". When she places her horn on the heart of another, she senses their future.

The name Ghel is an Indo-European word which means yellow. The word "gold" most likely has its origins in the word "Ghel".

Gold is known to possess spiritual powers that bring happiness, peace, stability and luck to those who wear it. Scientists say that all the gold in the world comes from the collision of neutron stars.

Though most other magical unicorn elementals get along well with the gold horn unicorn herd; Ghel, and others like her, must be very careful around the Quick Silver Herd - as gold dissolves in mercury.

Do Your Middle Graders Want To Know More From The Magical Elementals About The Periodic Table?

Get the accompanying books in print at all online book stores. Get the books and accompanying activities at MagicalPTElements.

Available Now

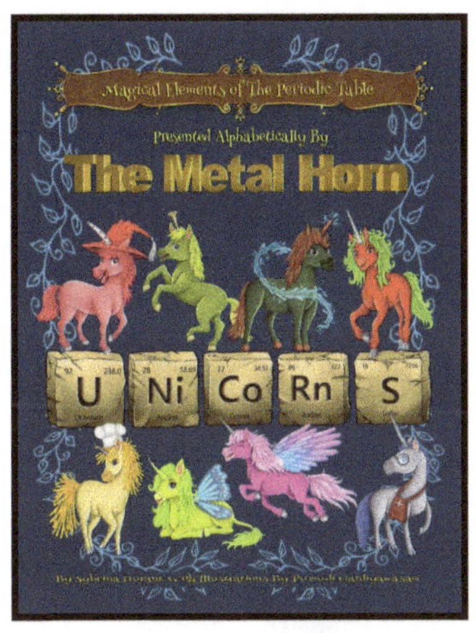

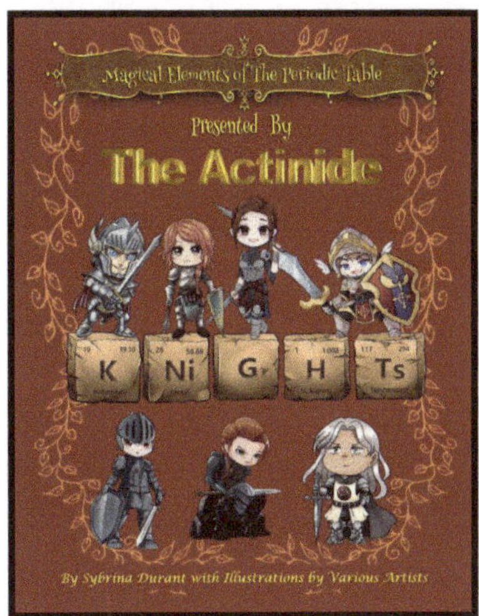

Learn More About all of the Periodic Table Elementals.
Get all of the "No Metal No Magic" Books Featuring Individual Elements at MagicalPTElements.com

Get These Fun Elemental Periodic Table Activities at

MagicalPTElements.

Unicorn Periodic Table Bingo—Comes with 32 unique Bingo cards. Magical Elementals Bingo comes with 36.

Magical Elemental Game Cards—Makes great prizes. Fun to trade, too.

1
2
3

plus

Unicorn **H**orn

Alphabet

CLIP ART FOR YOUR GRAPHICS

A
B
C

Also browse activities at
https://www.magicalPTelements.com
for all kinds of printable downloads to make learning fun.

Printable Magical Elemental Activity Downloads

Fun Way For Students To Learn The Elements Of The Periodic Table

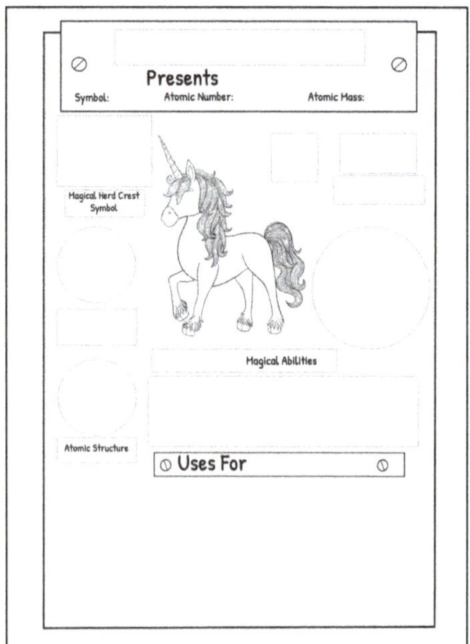

Blank Unicorn Element Card

Sample Unicorn Element Card

Blank Research Sheet

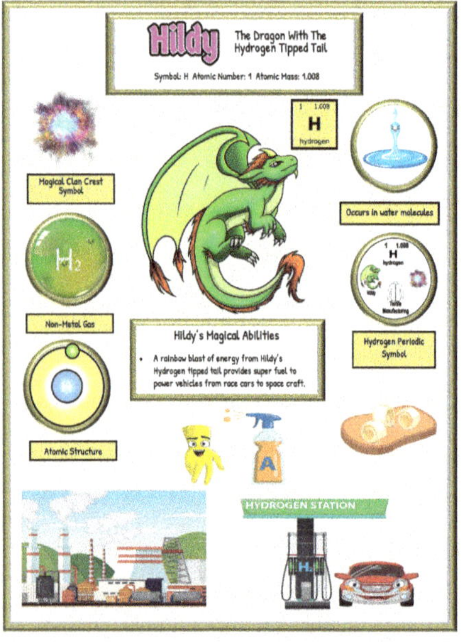

Sample Dragon Element Card

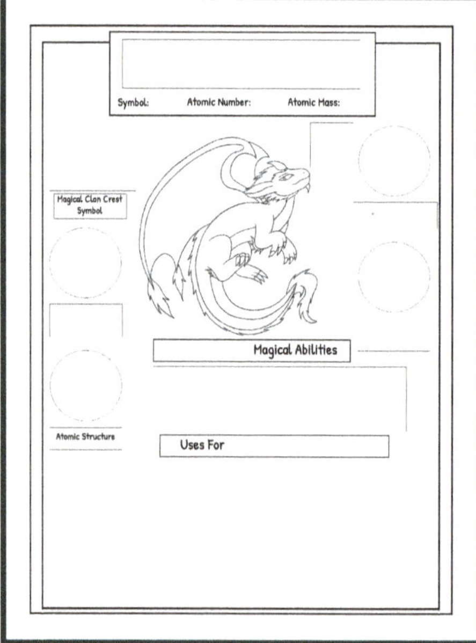

Blank Dragon Element Card

Blank Research Sheet

Using the sample Magical Elemental cards provided, have students select an element from the Periodic Table and a Magical Elemental Card Blank to create their own Magical Elemental Card. The blank and sample cards do not have to match.

You will receive a pdf containing either 26 unicorn or 26 dragon sample cards and blanks to be printed on 8 1/2 x 11 sized paper or card stock. The pdf also contains a Magical Elemental Research Sheet for the students to work on before creating their unique Periodic Table Elemental. They will also write a short paragraph describing their Unicorn or Dragon Elemental from that research.

Get These Fun Elemental Periodic Table Activity Sheets at *MagicalPTElements.com*

Don't Forget To Get A Tee Shirt

Featuring Your Favorite of the

118 Elements from the Periodic Table

Available in adult and kid sizes in many colors.

https://amzn.to/47NVZWN

This is the Hydrogen-Hildy Tee Shirt Graphic

Get it at https://www.amazon.com/dp/B0D3Y64G8T

Dear Reader

I hope the "No Metal No Magic Element 1 — Hydrogen, Presented By Hildy, from The Magical Elements of the Periodic Table Book Series" with illustrations by Pranavva et al, has helped you learn some fun and interesting things about the magic of the element, Hydrogen.

This is one of what will eventually be 118 books featuring periodic table elements presented by unicorns, dragons, wizards, knights and goblins. Keep checking regularly. Every one of the elements are amazing and very necessary to our

Techno-magical.

A lot of research went into every page of this book as well as the Magical Elements of the Periodic Table Books. There are just too many references to publish in this book but you can read and research them all at MagicalPTElements.com/MAUPT or /MDAPT or /MW1PT or / MW2PT or /MAKPT or /MRGPT. There, you can also access book related activity sheets and games to help make the learning process more fun.

Get ready made trading cards, lapel pins, tee shirts and more based on this book from Sybrina Publishing's No Metal No Magic Collection at Zazzle - **http://bit.ly/3km64Wg**

Would you like a 24" x 36" poster of the Elemental-Themed Periodic Table in this book? The best place to get it is at **https://bit.ly/49QMxBT** They have the sharpest images of any other poster printer around.

The Magical Elements of the Periodic Table books came into existence because of my Blue Unicorn—Journey To Osm books. If it weren't for their magical powers, based on the properties of the metals of their horns and hooves, I would have never come up with the idea to relate magical creatures to the periodic table. There's a metal horn unicorn story for every age group and they are all available at MagicalPTElements.com

If you enjoyed this book
please leave a nice review
at your favorite online book site.

No Metal No Magic

Song Lyrics

No metal, no Magic

No metal, no Magic

I can think of nothing more tragic

Than to have no metal or no magic

Metal makes everything magical.

Just ask a unicorn. . .

Preferably, one with a metal horn.

They'd say No metal, No magic.

Metal makes everything techno magical.

No metal, No magic

for two-leggers or unicorns.

No metal, No magic

Metal makes everything techno magical.

No metal, No magic

It's techno magical.

No metal, No magic

It might be very hard to believe but with

No metal, No magic

There'd be no technology.

No metal, No magic

Listen to this song at https://youtu.be/tcB8KDWAd8w

Watch the book trailer at https://youtu.be/NIX9fE7GJRI

Blue Unicorn

Ebooks, Audio and Print Books Available at all online book stores.

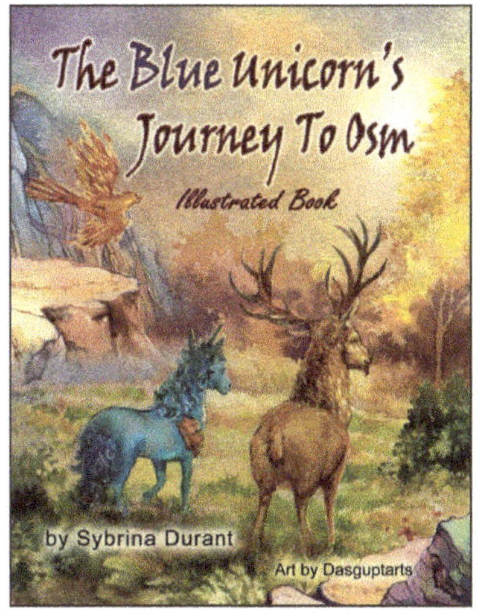

**Illustrated
Book**

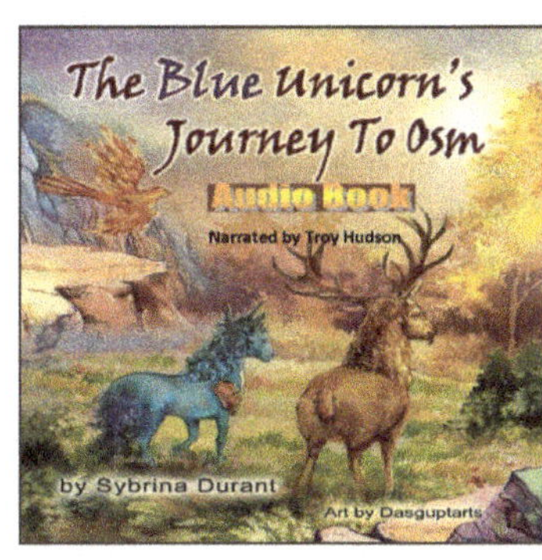

**Audio
Book**

**'Read & Color' Book
For Teens**

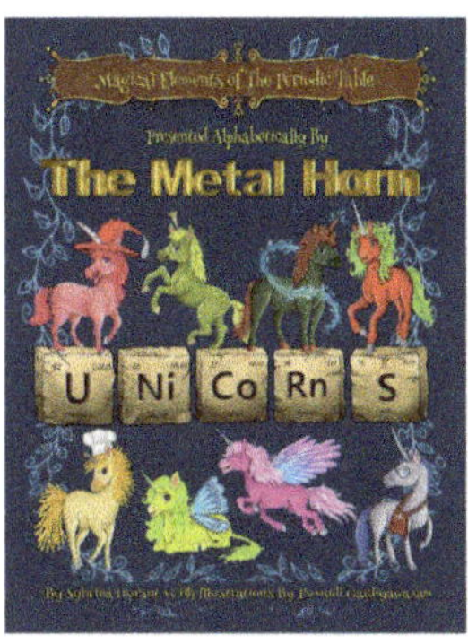

**Unicorn Periodic
Table Book**

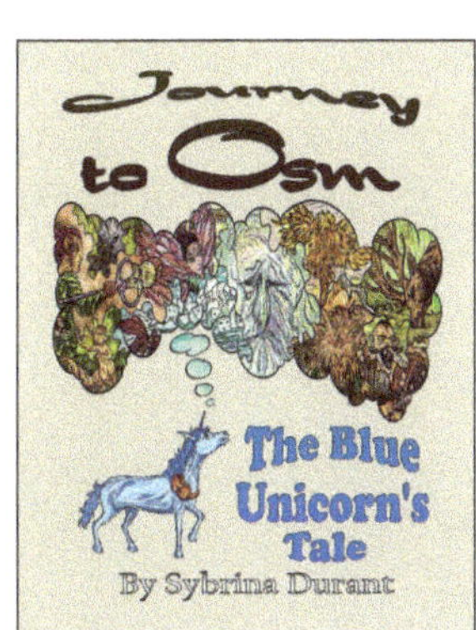

**Fantasy
Novel**

**Coloring Book & Character
Introduction**

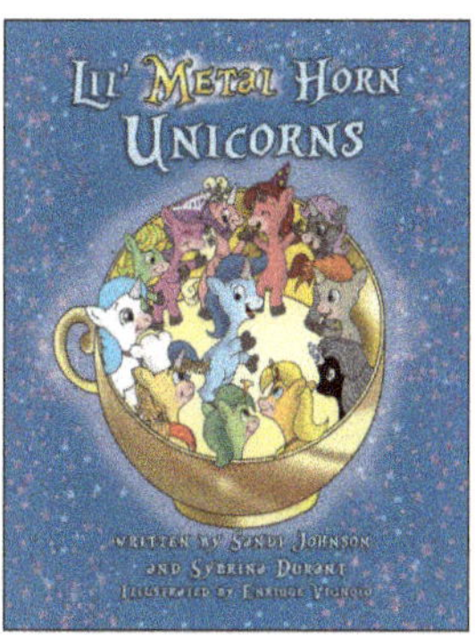

**Picture Book For
Kids**

Get these and more at
MagicalPTElements.com
and Sybrina.com

Back then, most places throughout MarBryn had wizards and sorcerers of some ability, or another. Most were trained in the ways of magical arts by the unicorns as part of their outreach program. Some two-leggers developed practical magical skills like making delicious feasts of tasty food appear out of thin air or purifying murky water around the land.

Others went through more extensive training to learn battle magic—like shooting powerful streams of energy from their swords.

Some were taught the art of holding the glow of the sun in magical globes, bringing light into the dark of night. These magical lights warded off the evil beings that were new and frightening products of dark magic.

With the rise of the sorcerer Magh, magical defense arts had become more important. The highest level of magical training involved sensing when others were in danger and learning to see into the future. Very few two-leggers ever reached that level because magic wasn't inherent in them the way it was for the unicorns. The metal of their horns and hooves were part of them as well as the very makeup of their blood, but two-leggers relied on learned magic via potions, charms, and incantations that required help from ingredients and forces more mystical in nature than any two-legger was ever born to be. Of course, controlling magic and projecting your intentions went far beyond merely following a recipe of sorts. It took being in touch with nature and the various elements to get the response a wizard desired. Much trial and error went into it, as well as faith and trust and the motives of the spell caster.

Magic was and is a practice that is never quite perfected even for the unicorns who must continue to hone and learn how to harness their powers. On rare occasion a wizard and unicorn had formed enough trust and a steadfast bond that prompted the unicorn to gift the wizard with a wand or staff embedded with the smallest sliver of metal from one of their hooves. This was rare but had happened and of course so had the desire for more power. Magh wasn't the first sorcerer with lust for more and throughout history there had been a handful of heinous acts against unicorns from those seeking their magic. Prior to Magh those wizards had failed to circumvent the protections nature had infused unicorn magic with so even after harvesting metal from their horns or hooves these sorcerers had gone mad trying to bend the will of nature and actually use their ill-gotten gains.

Magh, too, had gone mad or perhaps he'd already been so, but somehow he'd managed to harness the metal he harvested from his victims and continued to grow stronger rather than completely lose his mind like the others. How exactly remained a mystery to the unicorns and everyone else.

The wizards who'd honed their craft with the help and blessing of the unicorns tried to solve the riddle and even the best of them failed to uncover that secret. Some of MarBryn's natives took to magic naturally, while others struggled with the concept but no matter how adept they were. All of them fought valiantly against Magh's magic because with even a shred of knowledge they understood how the power shift would ultimately play out. They fought to the end but, in the end, only one sorcerer remained in MarBryn and now, Magh was in total control. But still, he was not satisfied. He wanted to control every living creature in the land.

Get The Novel at MagicalPTElements.com